Classical Mechanics

This book of problems and solutions in classical mechanics is dedicated to junior or senior undergraduate students in physics, engineering, applied mathematics, astronomy, or chemistry who may want to improve their problem-solving skills, or to freshman graduate students who may be seeking a refresh of the material.

The book is structured in 10 chapters, starting with Newton's Laws, motion with air resistance, conservation laws, oscillations, and the Lagrangian and Hamiltonian Formalisms. The last two chapters introduce some ideas in nonlinear dynamics, chaos, and special relativity. Each chapter starts with a brief theoretical outline and continues with problems and detailed solutions. A concise presentation of differential equations can be found in the appendix. A variety of problems are presented, from the standard classical mechanics problems, to context-rich problems and more challenging problems.

Key Features:

- Presents a theoretical outline for each chapter
- Motivates the students with standard mechanics problems with step-by-step explanations
- Challenges the students with more complex problems with detailed solutions

Classical Mechanics

Problems and Solutions

Carolina C. Ilie
Zachariah S. Schrecengost
Elina M. van Kempen

CRC Press
Taylor & Francis Group
Boca Raton London New York

CRC Press is an imprint of the
Taylor & Francis Group, an **informa** business

First edition published 2023
by CRC Press
4 Park Square, Milton Park, Abingdon, Oxon, OX14 4RN

and by CRC Press
6000 Broken Sound Parkway NW, Suite 300, Boca Raton, FL 33487-2742

© 2023 Taylor & Francis Group, LLC

CRC Press is an imprint of Informa UK Limited

British Library Cataloguing-in-Publication Data
A catalogue record for this book is available from the British Library

ISBN: 978-0-367-76844-7 (hbk)
ISBN: 978-1-032-43099-7 (pbk)
ISBN: 978-1-003-36570-9 (ebk)

DOI: 10.1201/9781003365709

Typeset in Times
by SPi Technologies India Pvt Ltd (Straive)

To my students: be extraordinary!!
To my family, mentors, and friends: thank you for the journey!!
-CCI

To my family, friends, and mentors.
-ZSS

To my family and SUNY Oswego friends.
-Elina van Kempen

Contents

Preface

This book of problems and solutions is written for undergraduate students in physics, mechanical engineering, applied mathematics, and chemistry, who may want to improve their skills in solving classical mechanics problems, or for first-year graduate students who may need a refresher. For a comprehensive textbook, the authors recommend "Classical Mechanics" by John R. Taylor, and for graduate level, "Classical Mechanics" by Herbert Goldstein.

The book is structured into ten chapters. The first eight chapters span the traditional spectrum of classical mechanics: starting with Newton's Laws, conservation laws, and oscillations, and ending with Lagrangian and Hamiltonian Formalisms, along with coupled oscillations. The last two chapters introduce some ideas in nonlinear dynamics, chaos, and special relativity. Each chapter starts with a brief theoretical outline.

The problems are organized from simple, standard pedagogical problems (e.g., Atwood machine, free fall, etc.) useful for beginners, and solved in great detail, to more involved problems, and each chapter ends with more challenging, interesting problems. The authors tried to keep a balance between simpler problems, context-rich problems, and more complex problems.

Good problem-solving skills can be acquired by following a few steps – becoming familiar with the theory and the notations, making sure the mathematical background is solid, then solving the problems without checking the solution, and finally, solution check. For solidifying the material, it may be useful to work on the problems in a few days and gradually add more difficult problems.

Acknowledgments

We thank Dr. Ilie's classical mechanics students for pointing out some of the more challenging concepts and motivating us to write this book. We are especially grateful to our illustrator, Dr. Julia R. D'Rozario, for creating the figures during a very busy time just before and after her Ph.D. defense. We thank Victor Sabirianov, who volunteered to do the final figure editing. Elina van Kempen thanks Mathieu van Kempen for his efficient and valuable help with proofreading and typing. Dr. Ilie would like to thank the wonderful co-authors Zachariah Schrecengost and Elina van Kempen for becoming a great, efficient, and fun writing team; working together was exciting and inspiring! Dr. Ilie is grateful to Dr. Peter Dowben, from the University of Nebraska at Lincoln for brilliant mentoring, summer research opportunities, coffee, and for inspiring her to write books. Many thanks to the administration at SUNY Oswego for overall support and for recognizing students and faculty scholarly and creative endeavors. Many thanks to our kind and professional editors, Dr. Danny Kielty, Rebecca Hodges-Davies, project manager Thivya Vasudevan (Straive), production editor Kari Budyk, and the full editing team from CRC Press and Taylor & Francis Group.

The authors thank our readers for their curiosity and joy in learning and encourage them to take time to enjoy challenge and discovery. Last, but not least, the authors would like to thank their families and friends for their love, joy, and funny jokes, which made this journey exciting.

About the Authors

Carolina C. Ilie is a Sigma Xi Fellow, Full Professor with tenure at the State University of New York at Oswego. She taught Classical Mechanics for more than ten years and she designed various problems for her students. Dr. Ilie obtained her Ph.D. in Physics and Astronomy at the University of Nebraska at Lincoln, an M.Sc. in Physics at the Ohio State University, and another M.Sc. in Physics at the University of Bucharest, Romania. She received the President's Award for Teaching Excellence in 2016 and the Provost Award for Mentoring in Scholarly and Creative Activity in 2013. Her research is focused on condensed matter physics: dynamics at surfaces, capillary condensation, perovskites photovoltaics. She lives in Central New York with her spouse, also a physicist, and their two sons. Her hobbies are classical music, learning languages, and writing.

Zachariah S. Schrecengost is a State University of New York alumnus. He graduated summa cum laude with a B.S. degree having completed majors in Physics, Software Engineering, and Applied Mathematics. He took the Advanced Mechanics course with Dr. Ilie and loved to be involved in this project. He brings to the project the experience of writing other two books of problems, but also the fresh perspective of the student taking classical mechanics and the enthusiasm and talent of an alumnus who is a physics and upper-level mathematics aficionado. Mr. Schrecengost works as software engineer in Syracuse while working toward his Ph.D. in Physics at Syracuse University.

Elina M. van Kempen graduated summa cum laude from the State University of New York at Oswego, with a double major in Physics and Applied Mathematics, and a minor in Computer Science. She is now working on her Ph.D. in Computer Science at the University of California, Irvine, with a focus on Security and Privacy. At SUNY Oswego, she had the great opportunity to take several courses taught by Dr. Carolina Ilie, including her course on Advanced Mechanics. Elina M. van Kempen also tutored for three years at SUNY Oswego, in Physics, Mathematics, and Computer Science and enjoyed helping students to understand and succeed in their classes. She loves traveling, cooking, and swimming.

ILLUSTRATIONS

JULIA R. D'ROZARIO

Julia R. D'Rozario (*illustrator*) graduated with a Ph.D. in Microsystems Engineering at Rochester Institute of Technology in Rochester, NY, in May 2022. Her doctoral research includes the study of light manipulation in optoelectronic devices, including III–V compound semiconductor space photovoltaics and micrometer-scale LEDs to improve device performance. Dr. D'Rozario aims to utilize the skillset she developed throughout grad school to help innovate technological advancements in her future career.

Also by Ilie and Schrecengost, with illustrations by Julia D'Rozario:

1. Carolina C. Ilie, Zachariah S. Schrecengost, *Electromagnetism: Problems and Solutions*, Institute of Physics IOP Science, UK, Morgan & Claypool Publishers, CA, USA; November 2016; Online ISBN: 978-1-6817-4429-2 • Print ISBN: 978-1-6817-4428-5
2. Carolina C. Ilie, Zachariah S. Schrecengost, *Electrodynamics: Problems and Solutions*, Institute of Physics IOP Science, UK and Morgan and Claypool Publishers, CA, USA; May 2018; Online ISBN: 978-1-6817-4931-0 • Print ISBN: 978-1-6817-4928-0

1 Newton's Laws

1.1 THEORY

This chapter introduces the three Newton's Laws and problems solved in a system of coordinates appropriate for each case. The authors hope that this chapter will bridge the knowledge and skills acquired in lower-level classical mechanics courses and bring the refinement of differential equations in different systems of coordinates and a broader vision on problem solving.

1.1.1 VECTORS

1.1.1.1 Space and Time

The motion of an object is analyzed by placing the object in a system of coordinates with a clock measuring the time t. The system of coordinates is chosen depending on the type of motion: a cartesian system of coordinates (x, y, z), a two-dimensional polar one (r, ϕ), a cylindrical one (s, ϕ, z), or a spherical system of coordinates (r, θ, ϕ).

1.1.1.2 Position, Velocity, and Acceleration

For an object moving from the initial position $\vec{r}_0$ at time t_0 to the final position $\vec{r}$ at time t, the velocity (instantaneous) is defined as

$$\vec{v} = \frac{d\vec{r}}{dt} = \dot{\vec{r}}$$

and the acceleration (instantaneous) as

$$\vec{a} = \frac{d\vec{v}}{dt} = \frac{d}{dt}\left(\frac{d\vec{r}}{dt}\right) = \frac{d^2\vec{r}}{dt^2} = \ddot{\vec{r}}$$

1.1.1.3 Scalar (Dot) and Vector (Cross) Product

Given vectors $\vec{A} = A_x\hat{x} + A_y\hat{y} + A_z\hat{z}$ and $\vec{B} = B_x\hat{x} + B_y\hat{y} + B_z\hat{z}$

The scalar product is a scalar equal to

$$\vec{A} \cdot \vec{B} = A_x B_x + A_y B_y + A_z B_z = AB\cos\theta$$

DOI: 10.1201/9781003365709-1

1

The vector product is a vector equal to

$$\vec{A} \times \vec{B} = \begin{vmatrix} \hat{x} & \hat{y} & \hat{z} \\ A_x & A_y & A_z \\ B_x & B_y & B_z \end{vmatrix} \quad \text{with} \quad |\vec{A} \times \vec{B}| = AB\sin\theta$$

where $A = |\vec{A}| = \sqrt{A_x^2 + A_y^2 + A_z^2}$, $B = |\vec{B}| = \sqrt{B_x^2 + B_y^2 + B_z^2}$, and θ is the angle between $\vec{A}$ and $\vec{B}$.

1.1.1.4 Gradient

Given a scalar function T, the gradients for various coordinate systems are given below.

Gradient in cartesian coordinates (x, y, z):

$$\nabla T = \frac{\partial T}{\partial x}\,\hat{x} + \frac{\partial T}{\partial y}\,\hat{y} + \frac{\partial T}{\partial z}\,\hat{z}$$

Gradient in cylindrical coordinates (s, ϕ, z):

$$\nabla T = \frac{\partial T}{\partial s}\,\hat{s} + \frac{1}{s}\frac{\partial T}{\partial \phi}\,\hat{\phi} + \frac{\partial T}{\partial z}\,\hat{z}$$

Gradient in spherical coordinates (r, θ, ϕ):

$$\nabla T = \frac{\partial T}{\partial r}\,\hat{r} + \frac{1}{r}\frac{\partial T}{\partial \theta}\,\hat{\theta} + \frac{1}{r\sin\theta}\frac{\partial T}{\partial \phi}\,\hat{\phi}$$

1.1.1.5 Line Integral

Given vector function $\vec{v}$ and path $\mathcal{P}$, a line integral is given by

$$\int_{\vec{a}\,\mathcal{P}}^{\vec{b}} \vec{v} \cdot d\vec{\ell}$$

where $\vec{a}$ and $\vec{b}$ are the end points and $d\vec{\ell}$ is the infinitesimal displacement vector along path $\mathcal{P}$. In Cartesian coordinates $d\vec{\ell} = dx\,\hat{x} + dy\,\hat{y} + dz\,\hat{z}$.

1.1.1.6 Surface Integral

Given vector function $\vec{v}$ and surface $\mathcal{S}$, a surface integral is given by

$$\int_{S} \vec{v} \cdot d\vec{a}$$

where $d\vec{a}$ is the infinitesimal area vector with direction normal to the surface. Note that $d\vec{a}$ always depends on the surface involved.

1.1.1.7 Volume Integral

Given scalar function T and volume $\mathcal{V}$, a volume integral is given by

$$\int_{\mathcal{V}} T\, dV$$

where dV is the infinitesimal volume element. In Cartesian coordinates $dV = dx\, dy\, dz$.

1.1.1.8 Kronecker Delta

$$\delta_{ij} = \begin{cases} 0 & \text{if } i \neq j \\ 1 & \text{if } i = j \end{cases}$$

For example,

$$\vec{A} \cdot \vec{B} = \sum_{i,j=1}^{n} A_i \delta_{ij} B_j = \sum_{i=1}^{n} A_i B_i$$

Levi-Civita symbol in three dimensions,

$$\varepsilon_{ijk} = \begin{cases} +1 \text{ if } (i,j,k) \text{ is an even permutation of } (1,2,3), \textit{i.e.,} (1,2,3), (2,3,1), \text{ or } (3,1,2) \\ -1 \text{ if } (i,j,k) \text{ is an odd permutation of } (1,2,3), \textit{i.e.,} (1,3,2), (2,1,3), \text{ or } (3,2,1) \\ 0 \text{ if } i=j, \text{ or } j=k, \text{ or } k=i \end{cases}$$

In three dimensions, Levi-Civita is related to the Kronecker delta in the "contracted epsilon identity"

$$\sum_{i=1}^{3} \varepsilon_{ijk}\varepsilon_{imn} = \delta_{jm}\delta_{kn} - \delta_{jn}\delta_{km}$$

1.1.2 COORDINATE SYSTEMS

1.1.2.1 Cartesian Coordinates

Here (Figure 1.1), our infinitesimal quantities are

$$d\vec{\ell} = dx\,\hat{x} + dy\,\hat{y} + dz\,\hat{z}$$

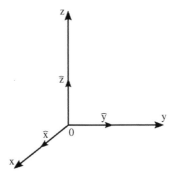

FIGURE 1.1   Cartesian system of coordinates.

The element of volume is expressed in the following way in Cartesian coordinates:

$$dV = dx\, dy\, dz$$

1.1.2.2   Cylindrical Polar Coordinates

Here (Figure 1.2), our infinitesimal quantities are

$$d\vec{\ell} = ds\,\hat{s} + s\,d\phi\,\hat{\phi} + dz\,\hat{z}$$

and element of volume

$$dV = s\, ds\, d\phi\, dz$$

1.1.2.3   Spherical Polar Coordinates

Here (Figure 1.3), our infinitesimal quantities are

$$d\vec{\ell} = dr\,\hat{r} + r\,d\theta\,\hat{\theta} + r\sin\theta\,d\phi\,\hat{\phi}$$

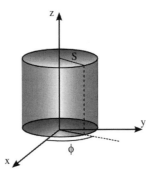

FIGURE 1.2   Cylindrical system of coordinates.

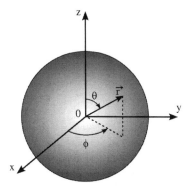

FIGURE 1.3   Spherical system of coordinates.

and the element of volume is

$$dV = r^2 \sin\theta\, dr\, d\theta\, d\phi$$

1.1.3   NEWTON'S LAWS

Isaac Newton was a student at Cambridge University when in 1665 plague swept through England. The university was closed and Newton decided to move back to his family's farm in Lincolnshire, where he lived for 18 months before moving back to Cambridge. That was for Newton his *annus mirabilis*, his miraculous year, when he discovered the nature of light, made a breakthrough in calculus more than ten years before Leibniz, and discovered gravity. Part of his results, what is known as Newton's Laws, was published in 1687 under the name *Principia Mathematica*.

1.1.3.1   Newton's First Law – The Law of Inertia

A particle maintains the state of rest or the state of motion with constant velocity as long as no net force is acting on it.

Inertia is the property of objects with mass to maintain their state of motion with constant velocity (the speed is constant and the direction is straight, so the velocity vector is constant), or their state of rest until an external force is acting on them to change this state.

1.1.3.2   Newton's Second Law

The acceleration of a particle is proportional to the net force acting on the particle and inversely proportional to the mass of the particle.

$$\vec{F} = m\vec{a}$$

$$\vec{a} = \frac{\vec{F}}{m}$$

In terms of linear momentum, $\vec{p} = m\vec{v}$.

$$\dot{\vec{p}} = m\dot{\vec{v}} = m\vec{a} \text{ (Here, it is assumed the mass is constant.)}$$

So, $\vec{F} = \dot{\vec{p}}$.

In classical mechanics, these two forms of Newton's Second Law are equivalent, with the second form more general than the first. However, in relativity, the derivative of momentum is different, because the mass is not constant at speeds close to the speed of light.

In different systems of coordinates, the Second Newton's Law is written in various forms.

In vector form,

$$\vec{F} = m\ddot{\vec{r}}$$

In cartesian coordinates (x, y, z),

$$F_x = m\ddot{x}$$

$$F_y = m\ddot{y}$$

$$F_z = m\ddot{z}$$

In two-dimensional polar coordinates (r, ϕ),

$$F_r = m(\ddot{r} - r\dot{\phi}^2)$$

$$F_\phi = m(r\ddot{\phi} + 2\dot{r}\dot{\phi})$$

In cylindrical coordinates (s, ϕ, z),

$$F_s = m(\ddot{s} - s\dot{\phi}^2)$$

$$F_\phi = m(s\ddot{\phi} + 2\dot{s}\dot{\phi})$$

$$F_z = m\ddot{z}$$

1.1.3.3 Newton's Third Law (Action–Reaction)

If an object 1 acts on object 2 with a force $\vec{F}_{12}$ (Figure 1.4), then the object 2 acts on object 1 with a force equal in magnitude, but in opposite direction: $\vec{F}_{21} = -\vec{F}_{12}$.

Note that in relativity, Newton's Third Law does not hold due to problems related to simultaneity.

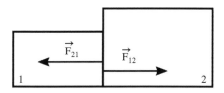

FIGURE 1.4 Two objects acting upon each other with forces equal in magnitude, but in opposite directions.

1.2 PROBLEMS AND SOLUTIONS

PROBLEM 1.1
Given vectors $\vec{A}$, $\vec{B}$, $\vec{C}$ prove the following vector relations using summation notation:

$$\vec{A} \cdot (\vec{B} \times \vec{C}) = \vec{C} \cdot (\vec{A} \times \vec{B})$$

$$\vec{A} \times (\vec{B} \times \vec{C}) = \vec{B}(\vec{A} \cdot \vec{C}) - \vec{C}(\vec{A} \cdot \vec{B})$$

SOLUTION 1.1

a. Looking at this in terms of components

$$\vec{A} \cdot (\vec{B} \times \vec{C}) = A_i \delta_{ij} (B_l C_m \varepsilon_{lmj}) = A_i B_l C_m \varepsilon_{lmi} = C_m A_i B_l \varepsilon_{ilm}$$
$$= C_k \delta_{kl} A_i B_l \varepsilon_{lmi} = \vec{C} \cdot (\vec{A} \times \vec{B})$$

as desired. Note there is a third relation which can be shown in a similar way, yielding

$$\vec{A} \cdot (\vec{B} \times \vec{C}) = \vec{C} \cdot (\vec{A} \times \vec{B}) = \vec{B} \cdot (\vec{C} \times \vec{A})$$

b. Looking at a particular component

$$[\vec{A} \times (\vec{B} \times \vec{C})]_k = A_i (B_l C_m \varepsilon_{lmj}) \varepsilon_{ijk} = A_i B_l C_m \varepsilon_{lmj} \varepsilon_{kij} = A_i B_l C_m (\delta_{lk}\delta_{mi} - \delta_{li}\delta_{mk})$$
$$= B_l \delta_{lk} (A_i C_m \delta_{mi}) - C_m \delta_{mk} (A_i B_l \delta_{li}) = B_k (A_i C_m \delta_{mi}) - C_k (A_i B_l \delta_{li})$$
$$= B_k (\vec{A} \cdot \vec{C}) - C_k (\vec{A} \cdot \vec{B}) = [\vec{B}(\vec{A} \cdot \vec{C}) - \vec{C}(\vec{A} \cdot \vec{B})]_k$$

Since this holds for all k, we have shown $\vec{A} \times (\vec{B} \times \vec{C}) = \vec{B}(\vec{A} \cdot \vec{C}) - \vec{C}(\vec{A} \cdot \vec{B})$.

PROBLEM 1.2

A mass m is hanging from a massless string of length R (Figure 1.5). The mass is moving in a circular motion with constant angular velocity ω. Using Newton's Second Law in polar coordinates, find the tension in the string.

SOLUTION 1.2

Newton's Second Law is applied (Figure 1.6):

$$\vec{F} = m\vec{a}$$

As specified, the polar components of acceleration, $a_r = \ddot{r} - r\dot{\phi}^2$ and $a_\theta = 2\dot{r}\dot{\theta} + r\ddot{\theta}$, are used. Thus,

$$F_r = m(\ddot{r} - r\dot{\phi}^2)$$

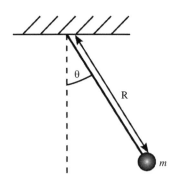

FIGURE 1.5   Mass hanging from the string.

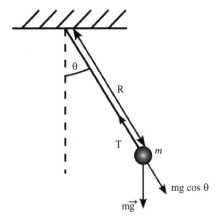

FIGURE 1.6   Mass hanging from the string, with gravitational force and tension in the string shown.

and

$$F_\theta = m(2\dot{r}\dot{\theta} + r\ddot{\theta})$$

As shown in Figure 1.6, the sum of the forces on the radial axis is

$$F_r = -T + mg\cos\theta$$

From this equation and Newton's Second Law, an expression for the tension in the string is obtained.

$$-T + mg\cos\theta = m(\ddot{r} - r\dot{\phi}^2)$$

Solving for T:

$$T = mg\cos\theta - m(\ddot{r} - r\dot{\phi}^2)$$

From the given information, this expression can be simplified. The mass is attached to a massless string of length R. Since the mass is always at distance R, it follows that and $r = R$ and $\dot{r} = \ddot{r} = 0$. It is also given that the mass moves with constant angular speed ω. Thus, $\dot{\phi} = \omega$.

The tension in the string is

$$T = m(g\cos\theta + R\omega^2)$$

PROBLEM 1.3

A rock of mass m falls down a cliff of height H (Figure 1.7). The rock has initial velocity v_0, directed horizontally. Assume no air resistance.

a. Find the position of the rock as a function of time.
b. What is the velocity of the rock when it touches the ground?

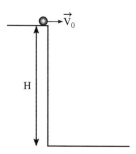

FIGURE 1.7 Rock on top of a cliff of height H.

SOLUTION 1.3

a. Newton's Second Law:

$$\vec{F} = m\vec{a}$$

In the x direction (Figure 1.8), no forces are acting on the rock

$$F_x = 0 = m\ddot{x}$$

In the y direction, the weight of the rock is considered

$$F_y = -mg = m\ddot{y}$$

First, by integrating $\int \ddot{x}\, dx = \int 0\, dt$, the x component of the rock's velocity, $\dot{x}$, is obtained:

$$\dot{x} = C_1 \tag{1.1}$$

At $t = 0$, $\dot{x}(0) = v_x(0) = v_0$. So $C_1 = v_0$. Now, solving for x is possible.

$$\int \dot{x}\, dx = \int v_0\, dt$$

$$x = v_0 t + C_2$$

At $t = 0$, $x(0) = 0$. So $C_2 = 0$. Solving for y as a function of time is done similarly by integrating.

$$\int m\ddot{y}\, dy = \int -mg\, dt$$

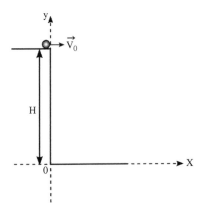

FIGURE 1.8   Rock on top of a cliff of height H.

$$\dot{y} = -gt + C_3 \tag{1.2}$$

At $t = 0$, $\dot{y}(0) = v_y(0) = 0$. So $C_3 = 0$.

$$\int \dot{y}\, dy = \int -gt\, dt$$

$$y = -\frac{gt^2}{2} + C_4$$

At $t = 0$, $y(0) = H$. So $C_4 = H$. The position of the rock as a function of time is described by

$$\begin{cases} x(t) = v_0 t \\ y(t) = -\dfrac{gt^2}{2} + H \end{cases}$$

b. The rock touches the ground at $y = 0$. First, the time at which the rock touches the ground is determined.

$$y(t_{\text{ground}}) = 0 = -\frac{gt_{\text{ground}}^2}{2} + H$$

$$t_{\text{ground}} = \sqrt{\frac{2H}{g}}$$

We substitute t_{ground} in Equations (1.1) and (1.2) and obtain the velocity of the rock when it reaches the ground:

$$\begin{cases} v_x(t_{\text{ground}}) = v_0 \\ v_y(t_{\text{ground}}) = -g\sqrt{\dfrac{2H}{g}} = -\sqrt{2gH} \end{cases}$$

PROBLEM 1.4

Consider a block of mass m sliding down a frictionless ramp at an incline θ (Figure 1.9). Find the velocity of the block at time t if the block is stationary at $t = 0$.

SOLUTION 1.4

To find the velocity, consider

$$\sum \vec{F}_{\text{ext}} = \dot{\vec{p}}$$

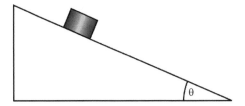

FIGURE 1.9   Block placed at the top of a ramp.

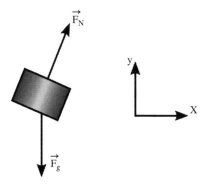

FIGURE 1.10   Forces on the block with respect to the coordinate system.

The coordinate system and forces can be taken as shown in Figure 1.10. The forces are then given by

$$\vec{F}_g = -mg\hat{y}$$

and

$$\vec{F}_N = mg(\sin\theta\,\hat{x} + \cos\theta\,\hat{y})\cos\theta$$

with

$$\dot{\vec{p}} = \vec{F}_N + \vec{F}_g = mg(\sin\theta\,\hat{x} + \cos\theta\,\hat{y})\cos\theta - mg\,\hat{y} = mg(\sin\theta\cos\theta\,\hat{x} + (\cos^2\theta - 1)\,\hat{y})$$

Using $\cos^2\theta + \sin^2\theta = 1$

$$\dot{\vec{p}} = mg\sin\theta\,(\cos\theta\,\hat{x} - \sin\theta\,\hat{y})$$

From this, the momentum is given by

$$\vec{p} = mgt\sin\theta\,(\cos\theta\,\hat{x} - \sin\theta\,\hat{y}) + \vec{C}$$

Since the velocity at $t = 0$ is zero, $\vec{p}(0) = 0$, which means $\vec{C} = 0$. Thus,

$$\vec{p} = mgt \sin\theta \, (\cos\theta \, \hat{x} - \sin\theta \, \hat{y})$$

From this, the velocity is given by

$$\vec{v} = \frac{\vec{p}}{m} = gt \sin\theta \, (\cos\theta \, \hat{x} - \sin\theta \, \hat{y})$$

the magnitude of which is

$$v = gt \sin\theta$$

as desired.

PROBLEM 1.5

A crow kicks a frictionless plastic plate (which does not rotate) with initial speed v_0, so that it slides straight up a snowy slippery roof that is inclined at an angle θ above the horizontal, then the crow flies to the plate and uses it as a sled down the roof.

    a. Write down Newton's Second Law for the lid and solve it to give its velocity as a function of time.
    b. Find the time needed for the lid to reach the highest point on the roof.
    c. Find how long the lid will take to return to its starting point, assuming the crow arrives to the lid exactly when it stops on the roof on its way up. Neglect friction and air drag.
    d. Find the position as a function of time.

SOLUTION 1.5

    a. The x-axis is chosen to be up the incline, as in the figure, and y-axis perpendicular to the plane (Figure 1.11).

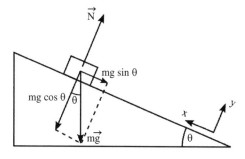

FIGURE 1.11 Inclined plane with the gravitational force decomposed on two perpendicular directions.

Newton's Second Law is applied in vectorial form and then written separately on each axis, noting that the acceleration on y-axis is zero.

$$\vec{F} = m\vec{a}$$

On y-axis, the resultant is zero (no acceleration).

$$F_y = N - mg\cos\theta = ma_y = 0$$

On x-axis, there is acceleration:

$$F_x = -mg\sin\theta$$

Also, $F_x = ma_x = m\dot{v}$. From both equations, and after dividing with the mass, it follows that:

$$\dot{v} = -g\sin\theta$$

Considering $\dot{v} = \dfrac{dv}{dt}$,

$$\frac{dv}{dt} = -g\sin\theta$$

By integration it follows that

$$\int dv = -\int g\sin\theta\, dt$$

From here it is easy to obtain the velocity as

$$v(t) = -gt\sin\theta + C$$

where C is the constant of integration to be obtained from the initial conditions when the clock starts: $t = 0\,s, v(t = 0) = v_0$. Therefore,

$$v(0) = v_0 = -g\cdot 0\cdot\sin\theta + C = C$$

and $C = v_0$. From here the velocity is

$$v(t) = v_0 - gt\sin\theta$$

It is important to note that, since the friction is neglected, the velocity is independent of mass.

b. In order to find the time t_{up} it is just needed to write that the velocity at that time will become zero, since the object stops.

$$0 = v_0 - g t_{\text{up}} \sin \theta$$

So indeed,

$$t_{\text{up}} = \frac{v_0}{g \sin \theta}$$

Also, the time t_{up} is mass independent.

c. Since there is no friction force and no air drag, the problem is symmetric (time upward is the same as time downward), and it is very easy to find the total time:

$$t_{\text{total}} = 2 \, t_{\text{up}}$$

Of course, there is another solution as well, by starting from the general formula for the velocity as a function of time:

$$v(t) = v_0 - g t \sin \theta$$

After the plate goes up and down (crow or no crow), the velocity on the way up will be of the same magnitude as the initial velocity, but in opposite direction $-v_0$ (here, symmetry is assumed as well), and the equation becomes

$$-v_0 = v_0 - g t_{\text{total}} \sin \theta$$

$$t_{\text{total}} = \frac{2 v_0}{g \sin \theta} = 2 t_{\text{up}}$$

$$t_{\text{total}} = 2 t_{\text{up}}$$

d. The position versus time is obtained by starting from the velocity formula $v(t) = \dfrac{dx(t)}{dt}$ and by integrating one more time.

$$v(t) = v_0 - g t \sin \theta$$

$$\frac{dx(t)}{dt} = v_0 - g t \sin \theta$$

so

$$\int dx = \int (v_0 - g t \sin \theta) \, dt$$

And the position as a function of time follows, with another constant of integration C'

$$x(t) = v_0 t - \frac{gt^2}{2} \sin\theta + C'$$

From the initial conditions, C' can be determined. At $t = 0$, $x = 0$

$$x(t = 0) = C' \text{ so } C' = 0$$

The position is

$$x(t) = v_0 t - \frac{gt^2}{2} \sin\theta$$

PROBLEM 1.6

A ball of mass m is sliding up a frictionless ramp of length l and at an angle θ. Determine the maximum initial velocity of the ball such that it doesn't slide off the ramp. Consider the following two cases and compare the answers:

a. The ball is initially released parallel to the ramp (Figure 1.12).
b. The ball is initially released parallel to the floor (Figure 1.13).

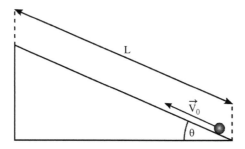

FIGURE 1.12   Ball traveling up frictionless ramp.

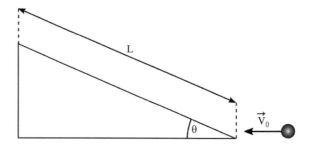

FIGURE 1.13   Ball traveling toward frictionless ramp.

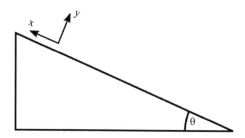

FIGURE 1.14 Coordinate system with respect to the ramp.

SOLUTION 1.6

a. The coordinate system is set up as shown in Figure 1.14.
So the initial velocity is given by

$$\vec{v}_0 = v_c \hat{x}$$

The force of gravity on the ball is given by

$$\vec{F}_g = -mg\sin\theta\,\hat{x} - mg\cos\theta\,\hat{y} = -mg(\sin\theta\,\hat{x} + \cos\theta\,\hat{y})$$

and the normal force is given by

$$\vec{F}_N = mg\cos\theta$$

Using these forces, in Newton's Second Law, the ball's acceleration can be determined

$$\sum \vec{F} = \vec{F}_g + \vec{F}_N = m\vec{a}$$

$$-mg(\sin\theta\,\hat{x} + \cos\theta\,\hat{y}) + mg\cos\theta\,\hat{y} = m\vec{a}$$

$$\vec{a} = -g\sin\theta\,\hat{x}$$

Since $\vec{a} = \dot{\vec{v}}$,

$$\dot{\vec{v}} = -g\sin\theta\,\hat{x}$$

By integration,

$$\vec{v} = -gt\sin\theta\,\hat{x} + \vec{C}_1$$

where $\vec{C}_1$ is some constant vector. Given that at $t = 0$, $\vec{v} = \vec{v}_0 = v_0\hat{x}$, one has

$$\vec{v} = (v_0 - gt\sin\theta)\hat{x}$$

Since it is required that $\vec{v} = 0$ at the top of the ramp, this takes

$$t = \frac{v_0}{g\sin\theta}$$

And $\vec{v} = \dot{\vec{r}}$, so

$$\dot{\vec{r}} = (v_0 - gt\sin\theta)\hat{x}$$

By integration,

$$\vec{r} = \vec{C}_2 + \left(v_0 t - \frac{1}{2}gt^2\sin\theta\right)\hat{x}$$

where $\vec{C}_2$ is another constant vector. The initial position can be taken to be zero so

$$\vec{r} = \left(v_0 t - \frac{1}{2}gt^2\sin\theta\right)\hat{x}$$

At the top of the ramp, $\vec{r} = L\hat{x}$. Since the time for the ball to reach the top of the ramp is known, everything can be plugged in and solved for v_0. Everything is in $\hat{x}$ so the unit vector can be dropped:

$$L = v_0 t - \frac{1}{2}gt^2\sin\theta$$

By substituting the time,

$$L = v_0\frac{v_0}{g\sin\theta} - \frac{1}{2}g\left(\frac{v_0}{g\sin\theta}\right)^2\sin\theta$$

$$L = \frac{v_0^2}{g\sin\theta} - \frac{1}{2}\frac{v_0^2}{g\sin\theta}$$

$$v_0 = \sqrt{2gL\sin\theta}$$

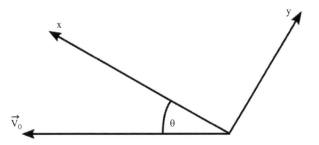

FIGURE 1.15   The initial velocity of the ball with respect to the coordinate system.

b. Now, the initial velocity is not parallel to the ramp. Keeping the same coordinate system as part a, the initial velocity is given in Figure 1.15. So

$$\vec{v}_0 = v_0 \cos\theta\,\hat{x} - v_0 \sin\theta\,\hat{y} = v_0(\sin\theta\,\hat{x} - \cos\theta\,\hat{y})$$

Since the ball does not move in the $\hat{y}$ direction once it is on the ramp, only the $\hat{x}$ part of the initial velocity must be considered. From part a, the constant $\vec{C}_1$ was just the initial velocity. Therefore, the velocity is then given by

$$\vec{v} = (v_0 \cos\theta - gt \sin\theta)\,\hat{x}$$

Again, it takes

$$t = \frac{v_0 \cos\theta}{g \sin\theta}$$

to reach the top. Also, suppose the ball hits the ramp at $t = 0$, then the initial position is 0 and

$$r = v_0 t \cos\theta - \frac{1}{2} gt^2 \sin\theta$$

At the top $\vec{r} = L\hat{x}$ and again, v_o can be solved for

$$L = \frac{v_0^2 \cos^2\theta}{g \sin\theta} - \frac{1}{2}\frac{v_0^2 \cos^2\theta}{g \sin\theta}$$

$$v_0 = \frac{1}{\cos\theta}\sqrt{2gL \sin\theta}$$

Note that as θ increases, the ball can be thrown faster and faster approaching $\theta = \frac{\pi}{2}$ where $v_0 \to \infty$ and the ball will never go past the ramp.

PROBLEM 1.7

A particle with initial mass m_0 and initial velocity v_0 begins losing mass according to the equation $m(t) = m_0 e^{-\alpha t}$ where α is constant. If there are no external forces, find an expression for the velocity.

SOLUTION 1.7

Since there are no external forces:

$$\vec{F}_{\text{ext}} = \dot{\vec{p}} = 0$$

Now that there is a changing mass, care must be taken with $\dot{\vec{p}}$.

$$\dot{\vec{p}} = \frac{d}{dt}(mv) = \dot{m}v + m\dot{v}$$

Therefore,

$$\dot{m}v + m\dot{v} = 0$$

Using $m(t)$ yields

$$-m_0\alpha e^{-\alpha t}v + m_0 e^{-\alpha t}\dot{v} = 0$$

After dividing both sides by $m_0 e^{-\alpha t}$,

$$\dot{v} - \alpha v = 0$$

$$\frac{dv}{dt} = \alpha v$$

Separation of variables,

$$\frac{dv}{v} = \alpha\, dt$$

$$\ln v = \alpha t + C$$

$$v(t) = Ae^{\alpha t}$$

where A and C are constants. Given $v(0) = v_0$, $A = v_0$. Therefore, the velocity is given by

$$v(t) = v_0 e^{\alpha t}$$

This makes sense: the particle exponentially loses mass so it must exponentially gain velocity. Note as $m \rightarrow 0$, $v \rightarrow \infty$, which clearly cannot happen considering the velocity must be less than the speed of light. However, it is always assumed to be in a regime where $v \ll c$.

PROBLEM 1.8

A particle of mass m moves with a velocity $v(t) = v(-\sin(\omega t)\hat{x} + \cos(\omega t)\hat{y})$. Is there a net external force acting on the particle? If so, find it and comment.

SOLUTION 1.8

If there is a net external force acting on the particle, $\dot{\vec{p}} \neq 0$. The momentum is given by

$$\vec{p}(t) = mv(-\sin(\omega t)\hat{x} + \cos(\omega t)\hat{y})$$

so

$$\dot{\vec{p}}(t) = -m\omega v(\cos(\omega t)\hat{x} + \sin(\omega t)\hat{y})$$

Note $\dot{\vec{p}} \neq 0$ for all t so there $\underline{is}$ a net external force acting on the particle given by

$$\vec{F}_{ext} = -m\omega v(\cos(\omega t)\hat{x} + \sin(\omega t)\hat{y})$$

Consider $\vec{v} \cdot \vec{F}_{ext}$,

$$\vec{v} \cdot \vec{F}_{ext} = -m\omega v^2(-\sin(\omega t)\cos(\omega t) + \cos(\omega t)\sin(\omega t)) = 0$$

Therefore, $\vec{v}$ and $\vec{F}_{ext}$ are orthogonal and the external force drives the particle in a circular motion.

PROBLEM 1.9

A soccer ball is hit from the ground level at an angle θ and an initial velocity v_0. Discuss the motion, considering that the air resistance is negligible.

Obtain: (a) the position versus time, (b) the time necessary for the ball to hit the ground, (c) the maximum height. Also: (d) discuss for which initial angle is the range maximized, (e) determine the equation of the trajectory in xy plane, that is, find $y(x)$.

SOLUTION 1.9

a. The x direction is chosen on the horizontal and y direction on the vertical. The z direction is not interesting, as no forces or velocities act in that direction. Newton's Second Law is applied, $\vec{F} = m\ddot{\vec{r}}$.

The forces acting on the soccer ball are identified in each direction. No air resistance, therefore, $F_x = 0$. Gravitational force acts in y direction, and considering the y-axis upward,

$$F_y = -mg$$

so $\vec{F} = (0, -mg, 0)$, therefore, the acceleration is $\ddot{\vec{r}} = (0, -g, 0)$.

The initial conditions are as follows: the initial position is chosen at the origin of the Cartesian system of coordinates, so $\vec{x}_0 = (0, 0, 0)$, while the initial velocity is $\vec{v}_0 = (v_0 \cos\theta, v_0 \sin\theta, 0)$.

The next step is to integrate on x and y directions separately and, by considering the initial conditions, the velocities are obtained.

$$\dot{x} = \int \ddot{x}\, dt = 0 + v_{0x} = v_0 \cos\theta$$

$$\dot{y} = \int \ddot{y}\, dt = -\int g\, dt = -gt + v_{0y} = -gt + v_0 \sin\theta$$

By another integration,

$$x(t) = \int \dot{x}\, dt = \int v_0 \cos\theta\, dt = v_0 t \cos\theta + x_0 = v_0 t \cos\theta$$

Similarly,

$$y(t) = \int \dot{y}\, dt = \int (-gt + v_0 \sin\theta)\, dt = -\frac{gt^2}{2} + v_0 t \sin\theta + y_0 = -\frac{gt^2}{2} + v_0 t \sin\theta$$

b. The time necessary for the ball to hit the ground can be easily found by imposing the condition $y(t) = 0$. It follows that

$$-\frac{gt^2}{2} + v_0 t \sin\theta = 0$$

so

$$t\left(-\frac{gt}{2} + v_0 \sin\theta\right) = 0$$

and the two solutions are $t = 0$ at the origin (when the ball is hit) and $t = \dfrac{2v_0 \sin\theta}{g}$, which is the total falling time.

c. It is important to recall that the vertical velocity is zero at the maximum height, otherwise the ball will go even higher. This is used to obtain the maximum height.

$$\dot{y}(t) = -gt + v_0 \sin\theta$$

$$\dot{y}(t) = 0$$

$$-gt + v_0 \sin\theta = 0$$

so

$$t = \frac{v_0 \sin\theta}{g}$$

which is half of the total falling time, which is expected due to the symmetry of the problem. Back to the y component

$$y(t) = -\frac{gt^2}{2} + v_0 t \sin\theta$$

The time necessary for the ball to reach the highest point is substituted and the highest position is obtained.

$$y = \frac{v_0^2 \sin^2\theta}{2g}$$

Another way to find the maximum height is using Galileo's equation $v_y^2 = v_{0y}^2 - 2gy$, and by replacing $v_y = 0$, the maximum height it is obtained as $y = \frac{v_{0y}^2}{2g} = \frac{v_0^2 \sin^2\theta}{2g}$, as before.

d. The range is calculated first.

$$x(t) = v_0 t \cos\theta$$

and using the total time $t = \frac{2v_0 \sin\theta}{g}$, the range is given by

$$R(\theta) = v_0 \frac{2v_0 \sin\theta}{g} \cos\theta = \frac{v_0^2 2 \sin\theta \cos\theta}{g} = \frac{v_0^2}{g} \sin 2\theta$$

The range is maximum for a given initial speed when $\sin 2\theta$ is maximum, which happens when

$$\sin 2\theta = 1$$

so

$$\theta = 45°$$

e. In order to find the equation of motion, the time is eliminated between the two equations for x and y.

$$x(t) = v_0 t \cos\theta$$

$$y(t) = -\frac{gt^2}{2} + v_0 t \sin\theta$$

From the first equation, the time is, as before, $t = \dfrac{x}{v_0 \cos\theta}$ and after substitution in the second equation the position y as a function of x is obtained:

$$y(x) = -\frac{g}{2}\left(\frac{x}{v_0\cos\theta}\right)^2 + v_0 \frac{x}{v_0\cos\theta}\sin\theta$$

$$= -\frac{g}{2(v_0\cos\theta)^2}x^2 + x\tan\theta = -\frac{g}{2(v_{0x})^2}x^2 + x\tan\theta$$

The position y is quadratic in x, which is expected since the trajectory is a parabola. Here, the angle θ is the initial angle.

PROBLEM 1.10

A mass m is suspended vertically to a spring with constant k. It is released from rest at $y(0) = y_r$ at $t = 0$. Describe the motion of the mass.

SOLUTION 1.10

Using Newton's Second Law, $\vec{F} = m\ddot{r} = m\ddot{y}$. The forces acting on the mass m are $\vec{F} = mg - ky$. Thus,

$$m\ddot{y} = mg - ky$$

$$\ddot{y} + \frac{k}{m}y = g$$

The solution of this differential equation is of the form $y = y_n + y_p$, where y_n is a solution to $\ddot{y} + \dfrac{k}{m} y = 0$, and y_p is a function satisfying $\ddot{y} + \dfrac{k}{m} y = g$.

$y_p = \dfrac{mg}{k}$ is a function that satisfies $\ddot{y} + \dfrac{k}{m} y = g$. Solving for $\ddot{y} + \dfrac{k}{m} y = 0$,

$y_n = C_1 \cos\left(\sqrt{\dfrac{k}{m}} t\right) + C_2 \sin\left(\sqrt{\dfrac{k}{m}} t\right)$. So,

$$y(t) = C_1 \cos\left(\sqrt{\dfrac{k}{m}} t\right) + C_2 \sin\left(\sqrt{\dfrac{k}{m}} t\right) + \dfrac{mg}{k}$$

Since $y(0) = y_r$, $C_1 = y_r - \dfrac{mg}{k}$. Since $\dot{y}(0) = 0$, $C_2 = 0$. The equation of motion is

$$y(t) = \left(y_r - \dfrac{mg}{k}\right) \cos\left(\sqrt{\dfrac{k}{m}} t\right) + \dfrac{mg}{k}$$

Note: $\dfrac{mg}{k} = \dfrac{ky_0}{k} = y_0$ is the shifted equilibrium. If the origin is chosen to be y_0,

$y(t) = y_r \cos\left(\sqrt{\dfrac{k}{m}} t\right)$.

PROBLEM 1.11

A particle of mass m is moving at a velocity v_0. At $t = 0$, a force $F = -\alpha v^2$, where α is a constant, acts on the particle. Find the particle's momentum as a function of time.

SOLUTION 1.11

Using Newton's Second Law

$$\vec{F} = m\vec{a}$$

$$-\alpha v^2 = m\dot{v}$$

This can be solved for v:

$$-\dfrac{\alpha}{m} v^2 = \dfrac{dv}{dt}$$

Separation of variables yields

$$-\dfrac{\alpha}{m} dt = v^{-2} dv$$

and integration yields

$$v^{-1} = \frac{\alpha}{m}t + C$$

At $t = 0$, $v = v_0$, so

$$\frac{1}{v_0} = C$$

and

$$\frac{1}{v} = \frac{\alpha}{m}t + \frac{1}{v_0}$$

Therefore,

$$v = \frac{mv_o}{\alpha v_0 t + m}$$

and the momentum is given by

$$p(t) = \frac{m^2 v_o}{\alpha v_0 t + m}$$

PROBLEM 1.12
A child on a sled, of combined mass m, is going up a snowy incline at an angle θ above the horizontal (Figure 1.16). The coefficient of friction between the sled and the snow is μ. The child has an initial speed v_0, parallel to the incline.

a. Find the position as a function of time $\vec{r}$.
b. What is the maximum distance traveled upward by the child?

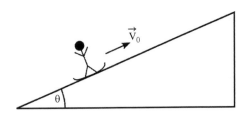

FIGURE 1.16   Child on a sled going up an incline.

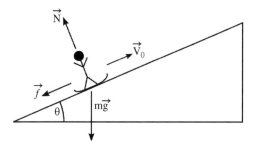

FIGURE 1.17   Child on a sled going up an incline.

SOLUTION 1.12

a. We choose the x-axis in the direction of the incline pointed upward (Figure 1.17). The y-axis is normal to the incline, pointed up. The z-axis is perpendicular to the x-axis and in the plane of the incline.
From Newton's Second Law, $\vec{F} = m\ddot{\vec{r}}$. Component-wise, we have

$$\begin{cases} F_x = m\ddot{x} \\ F_y = m\ddot{y} = 0 \\ F_z = m\ddot{z} = 0 \end{cases}$$

The forces acting on the system are

$$\begin{cases} F_x = -\mu N - mg \sin\theta \\ F_y = 0 \\ F_z = N - mg \cos\theta \end{cases}$$

Thus, $N - mg \cos\theta = 0$, so the value of N is $mg \cos\theta$.
From F_x, the following relationship is obtained:

$$m\ddot{x} = -\mu mg \cos\theta - mg \sin\theta$$

Integration is performed to get $\dot{x}$.

$$\dot{x} = \int -(\mu g \cos\theta + g \sin\theta)\, dt$$

$$\dot{x} = -(\mu g \cos\theta + g \sin\theta)t + C_1$$

At $t = 0$, $\dot{x}(0) = v_x(0) = v_0$. So $C_1 = v_0$. Integration is now performed to get x.

$$x = \int (-(\mu g \cos\theta + g \sin\theta)t + v_0)\, dt$$

$$x = -\frac{(\mu g \cos\theta + g\sin\theta)t^2}{2} + v_0 t + C_2$$

At $t = 0$, $x(0) = 0$. So $C_2 = 0$. The position as a function of time is

$$\vec{r} = \left(-\frac{(\mu g \cos\theta + g\sin\theta)t^2}{2} + v_0 t \right)\hat{x}$$

b. The child is at a maximum distance upward when $v(t) = \dot{x}(t) = 0$. The time t_{max} at which this maximum distance is reached is

$$-(\mu g \cos\theta + g\sin\theta)t + v_0 = 0$$

$$t_{max} = \frac{v_0}{\mu g \cos\theta + g\sin\theta}$$

Plugging this value into $x(t)$, the maximum distance is obtained:

$$x(t_{max}) = -\frac{(\mu g \cos\theta + g\sin\theta)}{2}\left(\frac{v_0}{\mu g \cos\theta + g\sin\theta}\right)^2 + v_0 \frac{v_0}{\mu g \cos\theta + g\sin\theta}$$

$$x(t_{max}) = \frac{v_0^2}{2(\mu g \cos\theta + g\sin\theta)}$$

PROBLEM 1.13

Describe the motion of a ladybug sitting on a vinyl disk of radius R rotating with a constant angular velocity ω in xy plane.

SOLUTION 1.13

The ladybug is at rest with respect to the disk, but in circular motion with respect to an observer. The trajectory is a circle of radius R and the ladybug has a constant angular velocity of ω. The linear velocity is perpendicular on the trajectory at every point (Figure 1.18) and has the magnitude $v = \omega R$.

The centripetal acceleration is oriented toward the center of the circle (from the Greek *centripetos* – toward the center) and has the magnitude of

$$a_{cp} = \frac{v^2}{R} = \omega^2 R$$

The position vector in xy plane is

$$\vec{r}(t) = R[\hat{x}\cos(\omega t) + \hat{y}\sin(\omega t)]$$

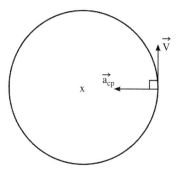

FIGURE 1.18 The velocity of the ladybug is tangent to the trajectory, while the centripetal acceleration is radial and perpendicular to the velocity.

The velocity can be found by differentiating the position vector with respect to time.

$$\dot{\vec{r}}(t) = \frac{d\vec{r}}{dt} = \omega R[-\hat{x}\sin(\omega t) + \hat{y}\cos(\omega t)]$$

The acceleration is obtained by differentiating again:

$$\ddot{\vec{r}}(t) = \frac{d\dot{\vec{r}}}{dt} = -\omega^2 R[\hat{x}\cos(\omega t) + \hat{y}\sin(\omega t)] = -\omega^2 \vec{r}(t)$$

Therefore, the acceleration is antiparallel (note the minus sign) to the radius and has indeed the magnitude

$$a_{cp} = \frac{v^2}{R} = \omega^2 R$$

PROBLEM 1.14

An unidentified object/particle of mass m has the following position as a function of time

$$\vec{r}(t) = e^{at}\,\hat{x} - \frac{1}{2}gt^2\,\hat{y}$$

where a is a constant and g is the gravitational acceleration. a) Find the velocity and the acceleration. b) Describe the force acting on this particle and describe how the motion depends on the constant a. c) Find the equation of the trajectory $y(x)$.

SOLUTION 1.14

a. The motion is two dimensional and the gravitational influence can be seen in y direction.

$$\vec{r}(t) = e^{at}\, \hat{x} - \frac{1}{2} g t^2\, \hat{y}$$

The velocity is easily obtained as

$$\vec{v}(t) = \frac{d\vec{r}}{dt} = \frac{d}{dt}\left(e^{at}\, \hat{x} - \frac{1}{2} g t^2\, \hat{y} \right) = \frac{d(e^{at})}{dt}\, \hat{x} - \frac{1}{2}\frac{d}{dt}(g t^2)\, \hat{y} = a e^{at}\, \hat{x} - g t\, \hat{y}$$

Similarly, the acceleration is obtained by one more differentiation as

$$\vec{a}(t) = \frac{d\vec{v}}{dt} = \frac{d(a e^{at})}{dt}\, \hat{x} - \frac{d(g t)}{dt}\, \hat{y} = a^2 e^{at}\, \hat{x} - g\, \hat{y}$$

b. The force acting on the object is

$$\vec{F}(t) = m\vec{a} = m(a^2 e^{at}\, \hat{x} - g\, \hat{y})$$

assuming that the mass is constant. The particle is moving in the negative y direction with gravitational acceleration and in the x direction with an acceleration which is exponentially decreasing in time if a is negative, and exponentially increasing in time if a is positive.

c. For the equation of the trajectory, let us write the position on each of the two directions:

$$x(t) = e^{at}$$

$$y(t) = -\frac{1}{2} g t^2$$

From the first equation, by applying the logarithm, it yields $\ln x = at$; therefore, $t = \frac{1}{a}\ln x$.

By substituting in the equation for y, it is obtained the equation of the trajectory as

$$y(x) = -\frac{g}{2a^2} \ln^2 x$$

The interesting thing about this fictious object/particle is that the motion is under sea level, so it may be a sea animal or a type of a submarine.

PROBLEM 1.15

A ball is placed at the top of a frictionless ramp with a height h and incline θ. The ball is also attached to a point $2h$ above the ground (h above the top of the ramp, as in Figure 1.19) with a string of length $2h$ and negligible mass. Find the initial velocity, down (parallel to) the ramp, the ball must have so that the highest position it reaches is halfway between the fixed point and the top of the ramp.

SOLUTION 1.15

The ball will slide down the ramp until it is a distance of $2h$ away from the fixed point (Figure 1.20). At this point it will travel upward, fixed in a circular arc due to the string. Consider the motion down the ramp and the pendulum motion separately.

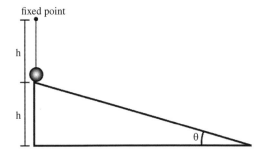

FIGURE 1.19   The ball positioned at the top of a ramp, connected to a fixed point.

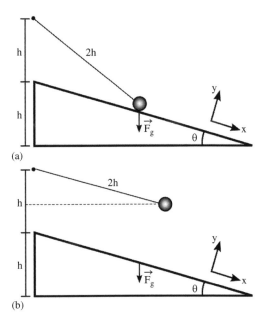

FIGURE 1.20   The ball moving as far down the ramp as possible before it must leave the surface of the ramp (a) and the ball in the final position in the air (b).

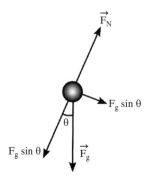

FIGURE 1.21   Forces acting on the ball.

First, it slides down the ramp with initial velocity v_{0x}.

The only forces on the ball are gravity and the normal force (Figure 1.21). During this part of the motion, there is no movement in the $\hat{y}$ direction, and the initial velocity is only in the $\hat{x}$ direction. Therefore, one can ignore the $\hat{y}$ part of the gravitational force and the normal force. Thus,

$$\Sigma F_x = ma_x$$

$$mg\sin\theta = ma_x$$

and the x component of acceleration is

$$a_x = g\sin\theta$$

Considering $\dot{\vec{v}} = \vec{a}$,

$$\dot{v}_x = g\sin\theta$$

$$v_x = v_{0x} + gt\sin\theta$$

where v_{0x} is what is ultimately solved for. In order to move to the next part of the calculation (the swinging portion of the movement), the ball's velocity the instant it leaves the ramp must be known. This requires the time the ball slides down the ramp. Considering $v_x = \dot{x}$, then

$$\dot{x} = v_{0x} + gt\sin\theta$$

By integration,

$$x = x_0 + v_{0x}t + \frac{1}{2}gt^2\sin\theta$$

Taking the initial position to be zero, the position is given by

$$x = v_{0x}t + \frac{1}{2}gt^2 \sin\theta$$

and solving for t yields

$$\frac{2x}{g\sin\theta} = \frac{2v_{0x}}{g\sin\theta}t + t^2$$

$$\left(t + \frac{v_{0x}}{g\sin\theta}\right)^2 = \frac{2x}{g\sin\theta} + \frac{v_{0x}^2}{g^2\sin^2\theta}$$

$$\left(t + \frac{v_{0x}}{g\sin\theta}\right)^2 = \frac{2xg\sin\theta + v_{0x}^2}{g^2\sin^2\theta}$$

$$t = \frac{\sqrt{2xg\sin\theta + v_{0x}^2} - v_{0x}}{g\sin\theta}$$

From the triangle in Figure 1.22:

$$(2h)^2 = h^2 + x^2 - 2hx\cos\left(\frac{\pi}{2} + \theta\right)$$

$$4h^2 = h^2 + x^2 + 2hx\sin\theta$$

$$3h^2 = x^2 + 2hx\sin\theta$$

$$(x + h\sin\theta)^2 = 3h^2 + h^2\sin^2\theta$$

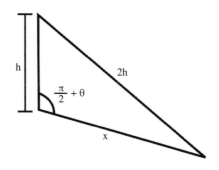

FIGURE 1.22 Geometry of the ball with respect to the ramp angle.

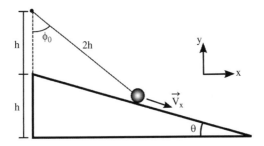

FIGURE 1.23   The velocity of the ball as it begins it swing motion.

$$x = h\left(\sqrt{\sin^2\theta + 3} - \sin\theta\right)$$

The velocity of the ball right as it loses contact with the ramp is

$$v_x = v_{0x} + gt\sin\theta = \sqrt{2xg\sin\theta + v_{0x}^2} = \sqrt{2gh\sin\theta\left(\sqrt{\sin^2\theta + 3} - \sin\theta\right) + v_{0x}^2}$$

For the swing part of the motion, take coordinates shown in Figure 1.23 with the fixed point as the origin.
The initial velocity is given by

$$\vec{v}_i = v_x(\cos\theta\,\hat{x} - \sin\theta\,\hat{y})$$

The force of gravity is now given by

$$\vec{F}_g = -mg\hat{y}$$

plus the tension in the string. Note, in switching to polar coordinates, everything in the $\hat{r}$ direction can be ignored since the motion is only in $\hat{\phi}$. Therefore,

$$\vec{F}_{tension} = \text{ignorable}$$

$$\vec{F}_g \rightarrow F_{g,\phi} = -mg\sin\phi$$

$$\vec{v}_i \rightarrow v_{i,\phi} = v_x(\cos\theta\cos\phi_0 - \sin\theta\sin\phi_0) = v_x\cos(\theta + \phi_0)$$

Newton's Second Law can now be used to find v_ϕ

$$F_{g,\phi} = ma_\phi = -mg\sin\phi$$

$$a_\phi = -g\sin\phi$$

Consider the following

$$\dot{v}_\phi = \frac{dv_\phi}{dt} = \frac{dv_\phi}{d\phi}\frac{d\phi}{dt} = \dot{\phi}\frac{dv_\phi}{d\phi}$$

It is also known that $v_\phi = 2h\dot{\phi}$. Thus,

$$\dot{v}_\phi = \frac{v_\phi}{2h}\frac{dv_\phi}{d\phi} = -g\sin\phi$$

Using separation of variables,

$$v_\phi dv_\phi = -2hg\sin\phi d\phi$$

$$\frac{1}{2}v_\phi^2 = 2hg\cos\phi + C$$

$$v_\phi^2 = 4hg\cos\phi + C$$

where C is a constant. Since the initial velocity is known, the initial angle is given as follows (Figure 1.24):

$$x^2 = h^2 + (2h)^2 - 2(2h)\cos\phi_0$$

$$h^2\left(\sqrt{\sin^2\theta + 3} - \sin\theta\right)^2 = 5h^2 - 4h^2\cos\phi_0$$

$$\cos\phi_0 = \frac{5 - \left(\sqrt{\sin^2\theta + 3} - \sin\theta\right)^2}{4}$$

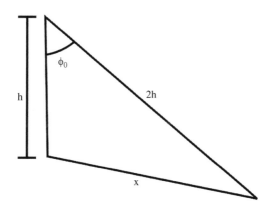

FIGURE 1.24   Geometry of the ball with respect to the fixed-point angle.

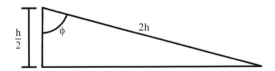

FIGURE 1.25   Geometry of the ball at the top of the swing motion.

Solving for C yields

$$v_x^2 \cos^2(\theta + \phi_0) = 4hg \cos \phi_0 + C$$

$$C = v_x^2 \cos^2(\theta + \phi_0) - 4hg \cos \phi_0$$

The velocity is now given by

$$v_\phi^2 = 4hg(\cos \phi - \cos \phi_0) + v_x^2 \cos^2(\theta + \phi_0)$$

When the ball reaches a maximum height halfway between the fixed point and the top of the ramp, $v_\phi = 0$ and ϕ is found as follows (Figure 1.25):

$$\phi = \cos^{-1} \frac{h}{4h} = \cos^{-1} \frac{1}{4}$$

Now, solving for v_x

$$0 = 4hg(\cos \phi - \cos \phi_0) + v_x^2 \cos^2(\theta + \phi_0)$$

$$v_x = \frac{2\sqrt{hg(\cos \phi_0 - \cos \phi)}}{\cos(\theta + \phi_0)}$$

Also,

$$\cos \phi_0 - \cos \phi = \frac{1}{4}\left(5 - \left(\sqrt{\sin^2 \theta + 3} - \sin \theta\right)^2 - 1\right) = \frac{1}{4}\left(\cos(2\theta) + 2\sqrt{\sin^2 \theta + 3}\right)$$

Therefore,

$$v_x = \frac{\sqrt{hg\left(\cos(2\theta) + 2\sqrt{\sin^2 \theta + 3}\right)}}{\cos\left(\theta + \cos^{-1} \dfrac{5 - \left(\sqrt{\sin^2 \theta + 3} - \sin \theta\right)^2}{4}\right)}$$

This can now be combined with the expression found for v_x before

$$v_x = \sqrt{2gh\sin\theta\left(\sqrt{\sin^2\theta+3}-\sin\theta\right)+v_{0x}^2}$$

Finally,

$$v_{0x} = \sqrt{\frac{hg\left(\cos(2\theta)+2\sqrt{\sin^2\theta+3}\right)}{\cos^2\left(\theta+\cos^{-1}\dfrac{5-\left(\sqrt{\sin^2\theta+3}-\sin\theta\right)^2}{4}\right)}-2gh\sin\theta\left(\sqrt{\sin^2\theta+3}-\sin\theta\right)}$$

PROBLEM 1.16

A plane is flying in the $\hat{x}$ direction at a velocity $\vec{v}_i = v_0\hat{x}$. It is delivering a package which is mysteriously attracted to the plane by the force $\vec{F} = \beta r\hat{r}$, where β is the strength of the attraction, r is the distance from the plane, in $\hat{r}$ points from the package to the plane. As soon as the package is dropped, the plane immediately begins traveling at v_f, 45° above $\hat{x}$. Find an expression for the height h the package should be dropped from such that its $\hat{y}$ velocity is zero when it hits the ground.

SOLUTION 1.16

The two forces acting on the package are gravity, $\vec{F}_g = -mg\hat{y}$, and the attractive force, $\vec{F} = \beta\vec{r}$, which can be written as

$$\vec{r} = \vec{r}_{\text{plane}} + \vec{r}_{\text{package}}$$

Take the origin of the coordinate system to be the spot where the package was released. Given the planes new velocity is $\vec{v}_{\text{plane}} = \dfrac{v_f}{\sqrt{2}}(\hat{x}+\hat{y})$, the position of the plane is

$$\vec{r}_{\text{plane}} = \frac{v_f t}{\sqrt{2}}(\hat{x}+\hat{y})$$

The position of the package can be expressed as

$$\vec{r}_{\text{package}} = x_p(t)\hat{x} + y_p(t)\hat{y}$$

Using Newton's Second Law yields

$$\sum \vec{F} = \beta \vec{r} - mg\hat{y} = \beta\left(\left(\frac{v_f t}{\sqrt{2}} - x_p\right)\hat{x} + \left(\frac{v_f t}{\sqrt{2}} - y_p\right)\hat{y}\right) - mg\hat{y}$$

$$= \beta\left(\frac{v_f t}{\sqrt{2}} - x_p\right)\hat{x} + \left(\beta\left(\frac{v_f t}{\sqrt{2}} - y_p\right) - mg\right)\hat{y} = m(a_x\hat{x} + a_y\hat{y})$$

Since the $\hat{y}$ velocity is the only one in question, the $\hat{x}$ velocity will be ignored. Therefore,

$$m\ddot{y}_p = \beta\frac{v_f t}{\sqrt{2}} - \beta y_p - mg$$

$$\ddot{y}_p + \frac{\beta}{m}y_p = \frac{\beta v_f}{\sqrt{2}}t - mg$$

Solving the homogeneous differential equation

$$\ddot{y}_p + \frac{\beta}{m}y_p = 0$$

yields

$$y_{p,h} = A\cos\left(\sqrt{\frac{\beta}{m}}t\right) + B\sin\left(\sqrt{\frac{\beta}{m}}t\right)$$

A particular solution can be obtained considering

$$y_{p,p} = Ct + D$$

where $\ddot{y} = 0$. Therefore,

$$\frac{\beta}{m}(Ct + D) = \frac{\beta}{m}\frac{v_f}{\sqrt{2}}t - g$$

with

$$C = \frac{v_f}{\sqrt{2}}$$

and

$$D = -\frac{mg}{\beta}$$

Combining these solutions yields

$$y_p = A\cos\left(\sqrt{\frac{\beta}{m}}t\right) + B\sin\left(\sqrt{\frac{\beta}{m}}t\right) + \frac{v_f}{\sqrt{2}}t - \frac{mg}{\beta}$$

It is known $y_p(0) = 0$, so

$$0 = A - \frac{mg}{\beta}$$

which yields A:

$$A = \frac{mg}{\beta}$$

and $\dot{y}_p(0) = 0$, so

$$0 = B\sqrt{\frac{\beta}{m}} + \frac{v_f}{\sqrt{2}}$$

and B is calculated to be

$$B = -v_f\sqrt{\frac{m}{2\beta}}$$

Finally,

$$y_p(t) = \frac{mg}{\beta}\left(\cos\left(\sqrt{\frac{\beta}{m}}t\right) - 1\right) + \frac{v_f}{\sqrt{2}}\left(t - \sqrt{\frac{m}{\beta}}\sin\left(\sqrt{\frac{\beta}{m}}t\right)\right)$$

Since it is required the velocity to be zero at the bottom of the motion, consider $\dot{y}_p$

$$\dot{y}_p = -g\sqrt{\frac{m}{\beta}}\sin\left(\sqrt{\frac{\beta}{m}}t\right) + \frac{v_f}{\sqrt{2}} - \frac{v_f}{\sqrt{2}}\cos\left(\sqrt{\frac{\beta}{m}}t\right)$$

Expressing $\sin\left(\sqrt{\dfrac{\beta}{m}}t\right)$ as s and $\cos\left(\sqrt{\dfrac{\beta}{m}}t\right)$ as c, this can be solved for c (and eventually t)

$$g\sqrt{\frac{m}{\beta}}s = \frac{v_f}{\sqrt{2}}(1-c)$$

$$\frac{mg^2}{\beta}s^2 = \frac{v_f^2}{2}(1-c)^2$$

After using $s^2 = 1 - c^2$

$$\frac{mg^2}{\beta}(1-c^2) = \frac{v_f^2}{2}(1-2c+c^2)$$

$$\left(\frac{v_f^2}{2}+\frac{g^2m}{\beta}\right)c^2 - v_f^2 c + \left(\frac{v_f^2}{2}-\frac{g^2m}{\beta}\right) = 0$$

Therefore,

$$c = \cos\left(\sqrt{\frac{\beta}{m}}t\right) = \frac{v_f^2 \pm \sqrt{v_f^4 - 4\left(\dfrac{v_f^2}{2}-\dfrac{g^2m}{\beta}\right)\left(\dfrac{v_f^2}{2}+\dfrac{g^2m}{\beta}\right)}}{2\left(\dfrac{v_f^2}{2}+\dfrac{g^2m}{\beta}\right)}$$

$$= \frac{v_f^2 \pm \sqrt{v_f^4 - 4\left(\dfrac{v_f^4}{4}-\dfrac{g^4m^2}{\beta^2}\right)}}{2\left(\dfrac{v_f^2}{2}+\dfrac{g^2m}{\beta}\right)}$$

Since $\cos\left(\sqrt{\dfrac{\beta}{m}}t\right)$ is bounded between negative one and one, the minus sign is taken. Therefore,

$$t = \sqrt{\frac{m}{\beta}}\cos^{-1}\frac{v_f^2 - \sqrt{v_f^4 - 4\left(\dfrac{v_f^4}{4}-\dfrac{g^4m^2}{\beta^2}\right)}}{2\left(\dfrac{v_f^2}{2}+\dfrac{g^2m}{\beta}\right)} + 2\pi n$$

for $n \in \mathbb{Z}$. Since h must be chosen such that $y_p(t) = -h$ where t is the time above, the package should be dropped from a height

$$h = \frac{mg}{\beta} \left(1 - \frac{v_f^2 - \sqrt{v_f^4 - 4\left(\dfrac{v_f^4}{4} - \dfrac{g^4 m^2}{\beta^2}\right)}}{2\left(\dfrac{v_f^2}{2} + \dfrac{g^2 m}{\beta}\right)} \right)$$

$$+ \frac{v_f}{\sqrt{2}} \sqrt{\frac{m}{\beta}} \sin\left(\cos^{-1} \frac{v_f^2 - \sqrt{v_f^4 - 4\left(\dfrac{v_f^4}{4} - \dfrac{g^4 m^2}{\beta^2}\right)}}{2\left(\dfrac{v_f^2}{2} + \dfrac{g^2 m}{\beta}\right)} \right)$$

$$- \sqrt{\frac{m}{\beta}} \cos^{-1} \frac{v_f^2 - \sqrt{v_f^4 - 4\left(\dfrac{v_f^4}{4} - \dfrac{g^4 m^2}{\beta^2}\right)}}{2\left(\dfrac{v_f^2}{2} + \dfrac{g^2 m}{\beta}\right)}$$

so that y velocity is zero when it lands.

2   Motion with Air Resistance

2.1   THEORY

This chapter introduces the motion of objects and projectiles with air resistance. Air resistance can be linear or quadratic, and in some circumstances, we need to consider both contributions.

2.1.1   DRAG FORCE OF AIR RESISTANCE

The drag force $\vec{f}$ has a magnitude dependent on speed, with its direction opposite to the velocity $\vec{v}$, as in Figure 2.1.

$$\vec{f} = -f(v)\hat{v}$$

For lower speeds (that is, much lower than the speed of light), the drag force has a linear term and a quadratic term, as follows

$$f(v) = f_{\text{linear}} + f_{\text{quadratic}} = bv + cv^2$$

with b and c constants with appropriate units.

The constants depend on the shape of the object, for example, for a spherical object, $b = \beta D$ and $c = \gamma D^2$, with D the diameter of the sphere and β and γ coefficients which depend on the medium.

The linear drag is due to the viscous drag of the medium (air, water, oil, etc.) and it depends on the viscosity of the medium and the linear size of the projectile. The quadratic term is proportional to the density of the medium and the cross-sectional area of the projectile. For some systems, the linear air drag can be neglected, for others, the quadratic air drag can be neglected, but in some circumstances, both linear and quadratic drag should be considered.

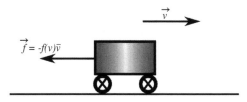

FIGURE 2.1   Cart moving to the right with velocity $\vec{v}$ under the influence of drag force $\vec{f}$.

DOI: 10.1201/9781003365709-2

2.2 PROBLEMS AND SOLUTIONS

PROBLEM 2.1

A bicycle is cruising in a horizontal motion with initial velocity v_0 under the influence of linear air drag. Analyze the motion, that is, find a) the velocity as a function of time and b) the position as a function of time. Neglect the friction forces.

SOLUTION 2.1

a. The net force is only the linear air drag, acting horizontally in a direction opposite to the velocity of the bicycle. The vertical forces – the gravitational force and the normal force of the surface – yield a zero vertical force, as in Figure 2.2.

$$\vec{F}_{net} = m\vec{g} + \vec{N} + \vec{f}_{linear} = \vec{f}_{linear} = -b\vec{v}$$

From Newton's Second Law, $\vec{F}_{net} = m\vec{a} = m\dot{\vec{v}}$, and considering only the x direction,

$$m\dot{v} = -bv$$

$$m\frac{dv}{dt} = -bv$$

Separation of variables,

$$\frac{dv}{v} = -\frac{b}{m}dt$$

The velocity is obtained by integrating on both sides the previous equation over velocity, from v_0 to v, and over time, from 0 to t.

$$\int_{v_0}^{v} \frac{dv'}{v'} = -\frac{b}{m}\int_0^t dt$$

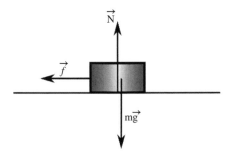

FIGURE 2.2 Forces acting on the bicycle.

And from here it yields

$$\ln v'\big|_{v_0}^{v} = -\frac{bt}{m}$$

Using the properties of the logarithm, $\ln A - \ln B = \ln \dfrac{A}{B}$, yields

$$\ln \frac{v}{v_0} = -\frac{bt}{m}$$

And from here, the speed as a function of time is calculated:

$$v(t) = v_0 e^{-\frac{bt}{m}}$$

Note that the velocity is exponentially decreasing to zero, reaching zero asymptotically as time approaches infinity.

b. The equation of motion can be obtained by integrating the speed obtained in part (a).

$$v = \frac{dx}{dt}$$

$$v(t) = v_0 e^{-\frac{bt}{m}}$$

$$dx = v_0 e^{-\frac{bt}{m}} \, dt$$

$$\int_{x_0}^{x} dx' = v_0 \int_{0}^{t} e^{-\frac{bt'}{m}} \, dt'$$

$$x - x_0 = -\frac{m}{b} v_0 e^{-\frac{bt'}{m}} \bigg|_0^t$$

And the position as a function of time is

$$x(t) = x_0 + \frac{m}{b} v_0 \left(1 - e^{-\frac{bt}{m}} \right)$$

As time approaches infinity, the object approaches a position equal to

$$x = x_0 + \frac{m}{b} v_0$$

PROBLEM 2.2

A projectile is falling into the atmosphere (gravitational acceleration g) with initial speed v_0 oriented vertically downward and it is subject to linear drag. a) Find the terminal velocity, v_{ter}. b) Obtain the speed at time t, $v(t)$.

SOLUTION 2.2

a. The two forces acting on the projectile are the gravitational force, downward, and the drag force upward (the direction opposite to the direction of the velocity, as in Figure 2.3).
   Newton's Second Law is written as

$$m\dot{v} = mg - bv$$

The terminal speed is the speed for which the drag force is balancing out the gravitational force, and therefore the acceleration becomes zero:

$$0 = mg - bv_{ter}$$

And the terminal speed for linear air drag is

$$v_{ter} = \frac{mg}{b}$$

b.

$$m\dot{v} = mg - bv$$

By using $v_{ter} = \dfrac{mg}{b}$, the previous equation becomes

$$m\dot{v} = b\frac{mg}{b} - bv$$

$$m\dot{v} = bv_{ter} - bv$$

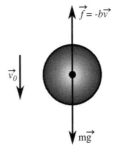

FIGURE 2.3   Projectile in vertical motion under the influence of gravitation and air drag.

The speed is obtained from

$$\frac{dv}{dt} = -\frac{b}{m}(v - v_{\text{ter}})$$

$$\frac{dv}{(v - v_{\text{ter}})} = -\frac{b}{m}dt$$

By integrating over speed from initial speed v_0 to v and over time from 0 to t

$$\int_{v_0}^{v} \frac{dv'}{(v' - v_{\text{ter}})} = -\int_{0}^{t} \frac{b}{m}dt'$$

$$\ln(v - v_{\text{ter}}) - \ln(v_0 - v_{\text{ter}}) = -\frac{b}{m}t$$

$$\ln\frac{(v - v_{\text{ter}})}{(v_0 - v_{\text{ter}})} = -\frac{b}{m}t$$

By using properties of logarithms and exponents, the speed as a function of time is obtained as

$$v(t) = v_{\text{ter}} + (v_0 - v_{\text{ter}})e^{-\frac{bt}{m}}$$

Introducing $\tau = \frac{m}{b}$, the speed becomes

$$v(t) = v_{\text{ter}} + (v_0 - v_{\text{ter}})e^{-\frac{t}{\tau}}$$

Or rearranging the terms

$$v(t) = v_{\text{ter}}\left(1 - e^{-\frac{t}{\tau}}\right) + v_0 e^{-\frac{t}{\tau}}$$

It is easy to see that, when the time goes to infinity, the speed goes asymptotically to terminal speed. If the projectile starts with zero initial velocity, then the speed becomes

$$v(t) = v_{\text{ter}}\left(1 - e^{-\frac{t}{\tau}}\right)$$

PROBLEM 2.3

A ball of mass m is thrown with initial velocity v_0 at an angle of 45° from the horizontal.

  a. Assuming no air resistance, find the trajectory of the ball $y(x)$.
  b. Assuming linear resistance bv, find the trajectory of the ball $y(x)$.

SOLUTION 2.3

  a. According to Newton's Second Law, $F_x = m\ddot{x} = 0$ and $F_y = m\ddot{y} = -mg$. First, an expression for x is computed by integrating twice.

$$\dot{x} = \int 0\, dt$$

$$\dot{x} = C_1$$

Since $\dot{x}(0) = v_{0x}$, $C_1 = v_{0x}$.

$$x = \int v_{0x}\, dt$$

$$x = v_{0x}t + C_2$$

Since $x(0) = 0$, $C_2 = 0$. So $x(t) = v_{0x}t$. Thus, $t = \dfrac{x}{v_{0x}}$.

Now, an expression for y is computed by integrating twice.

$$\dot{y} = \int -g\, dt$$

$$\dot{y} = -gt + C_3$$

Since $\dot{y}(0) = v_{0y}$, $C_3 = v_{0y}$.

$$y = \int -gt + v_{0y}\, dt$$

$$y = -\frac{1}{2}gt^2 + v_{0y}t + C_4$$

Since $y(0) = 0$, $C_4 = 0$. So $y(t) = -\dfrac{1}{2}gt^2 + v_{0y}t$.

By plugging t as a function of x in $y(t)$, a result for $y(x)$ is

$$y(x) = -\frac{gx^2}{2v_{0x}^2} + \frac{v_{0y}}{v_{0x}}x$$

b. The process is similar to part (a). According to Newton's Second Law, $\vec{F_x} = m\ddot{x} = -bv_x$ and $\vec{F_y} = m\ddot{y} = -mg - bv_y$. First, an expression for x is computed by integrating twice.

$$\int_{v_{0x}}^{v_x} \frac{dv'_x}{v'_x} = \int_0^t -\frac{b}{m} dt$$

$$\ln\left(\frac{v_x}{v_{0x}}\right) = -\frac{b}{m}t$$

$$v_x = v_{0x}e^{-\frac{b}{m}t}$$

After getting an expression for v_x, integration is performed a second time to solve for x.

$$x = \int_0^t v_{0x}e^{-\frac{b}{m}t'} dt'$$

$$x = \frac{v_{0x}m}{b}\left(1 - e^{-\frac{b}{m}t}\right)$$

Thus,

$$\left(1 - e^{-\frac{b}{m}t}\right) = \frac{bx}{v_{0x}m}$$

and

$$t = -\frac{m}{b}\ln\left(1 - \frac{bx}{v_{0x}m}\right)$$

Now, an expression for y is computed by integrating twice.

$$\int_{v_{0y}}^{v_y} \frac{dv'_y}{-g - \frac{b}{m}v'_y} = \int_0^t dt'$$

$$-\frac{m}{b}\ln(bv_y + mg) + \frac{m}{b}\ln(bv_{0y} + mg) = t$$

$$v_y = v_{0y}e^{-\frac{b}{m}t} - \frac{mg}{b}\left(1 - e^{-\frac{b}{m}t}\right)$$

After getting an expression for v_y, integration is performed a second time to solve for y.

$$y = \int_0^t \left(e^{-\frac{b}{m}t'}v_{0y} - \frac{gm}{b}\left(1 - e^{-\frac{b}{m}t'}\right) \right) dt'$$

$$y = \left(1 - e^{-\frac{b}{m}t}\right)\left(\frac{mv_{0y}}{b} + \frac{gm^2}{b^2} \right) - \frac{gm}{b}t$$

By plugging t as a function of x in $y(t)$, a result for $y(x)$ is

$$y(x) = \frac{xb}{v_{0x}m}\left(\frac{mv_{0y}}{b} + \frac{gm^2}{b^2} \right) - \frac{gm}{b}\left(-\frac{m}{b}\ln\left(1 - \frac{xb}{v_{0x}m}\right) \right)$$

$$y(x) = \frac{x}{v_{0x}}\left(v_{0y} + \frac{gm}{b} \right) + \frac{gm^2}{b^2}\ln\left(1 - \frac{bx}{v_{0x}m}\right)$$

PROBLEM 2.4

A bicycle is cruising in a horizontal motion with initial velocity v_0 under the influence of the quadratic air drag. Analyze the motion, that is, find a) the velocity as a function of time and b) the position as a function of time. Neglect the friction forces.

SOLUTION 2.4

a. In this case, the net force is the quadratic air drag. As expected, the gravitational force and the normal force of the surface yield a zero vertical force.

$$\vec{F}_{net} = m\vec{g} + \vec{N} + \vec{f}_{quadratic} = \vec{f}_{quadratic} = -cv^2\,\hat{v}$$

$$m\dot{v} = -cv^2$$

$$m\frac{dv}{dt} = -cv^2$$

$$\frac{dv}{v^2} = -\frac{c}{m}dt$$

The velocity is obtained by integrating on both sides the previous equation over velocity, from v_0 to v, and over time, from 0 to t.

$$\int_{v_0}^{v} \frac{dv'}{v'^2} = -\frac{c}{m} \int_{0}^{t} dt'$$

And from here it yields

$$-\frac{1}{v'}\Big|_{v_0}^{v} = -\frac{ct}{m}$$

$$\frac{1}{v} - \frac{1}{v_0} = \frac{ct}{m}$$

And from here we obtain the speed as a function of time:

$$v(t) = \frac{1}{\dfrac{1}{v_0} + \dfrac{ct}{m}} = \frac{v_0}{1+qt}$$

where $q = \dfrac{cv_0}{m}$. Note that the velocity is decreasing to zero as time goes to infinity.

b. The equation of motion can be obtained by integrating the speed obtained in part (a).

$$v = \frac{dx}{dt}$$

$$v(t) = \frac{v_0}{1+qt}$$

$$dx = \frac{v_0}{1+qt} dt$$

$$\int_{x_0}^{x} dx' = v_0 \int_{0}^{t} \frac{1}{1+qt'} dt'$$

$$x - x_0 = -\frac{v_0}{q} \ln(1+qt')\Big|_{0}^{t}$$

And the position as a function of time is

$$x(t) = x_0 + \frac{v_0}{q}(\ln(1+qt) - \ln 1) = x_0 + \frac{v_0}{q}\ln(1+qt)$$

PROBLEM 2.5

A bobsled of mass m is launched on a straight horizontal plane with initial speed v_0. It is under quadratic air resistance cv^2. What is the equation of position of the bobsled? Assume no friction.

SOLUTION 2.5

Using Newton's Second Law, $\vec{F_x} = m\ddot{x} = m\dot{v}$. Air resistance being the only force acting on the bobsled, it follows that $m\dot{v} = -cv^2$. By integrating, an expression for v is obtained.

$$\int_{v_0}^{v} \frac{dv'}{v'^2} = \int_{0}^{t} -\frac{c}{m} dt'$$

$$-\frac{1}{v} + \frac{1}{v_0} = -\frac{c}{m} t$$

$$v = \frac{1}{\dfrac{1}{v_0} + \dfrac{c}{m} t}$$

Since $v = dx/dt$, the position $x(t)$ of the bobsled is

$$x(t) = \int_{0}^{t} \frac{1}{\dfrac{1}{v_0} + \dfrac{c}{m} t'} dt'$$

$$x(t) = \frac{m}{c} (\ln(cv_0 t + m) - \ln(m))$$

PROBLEM 2.6

A projectile is falling in the atmosphere (gravitational acceleration g) with initial speed v_0 oriented vertically downward and is subject to quadratic drag. (a) Find the speed $v(t)$. (b) Obtain the equation of motion, that is, obtain $y(t)$.

SOLUTION 2.6

    a. In this case, the motion is constrained in the y direction, with the gravitational force oriented vertically downward and the quadratic drag upward. Newton's Second Law is written as

$$m\dot{v} = mg - cv^2$$

As before, the projectile will increase in speed, but the drag force will increase as well (quadratic) until gravitational force will balance out the drag force, and the acceleration becomes zero. The maximum speed is the terminal speed v_{ter}, which, as expected, will have a different form than in the linear drag case.

$$0 = mg - cv_{ter}^2$$

$$v_{ter} = \sqrt{\frac{mg}{c}}$$

Now, rewriting Newton's Second Law in terms of the speed and of the terminal speed, it follows

$$m\frac{dv}{dt} = c(v_{ter}^2 - v^2)$$

Generally, for equations of this type, it is a good option to express them in dimensionless quantities as follows:

$$m\frac{dv}{dt} = cv_{ter}^2\left(1 - \frac{v^2}{v_{ter}^2}\right)$$

$$\frac{dv}{\left(1 - \frac{v^2}{v_{ter}^2}\right)} = \frac{c}{m}v_{ter}^2 dt$$

By substituting terminal velocity on the right side

$$\frac{dv}{1 - \frac{v^2}{v_{ter}^2}} = g\, dt$$

By integration,

$$\int_{v_0}^{v}\frac{dv'}{1 - \frac{v'^2}{v_{ter}^2}} = g\int_0^t dt'$$

It yields that

$$v_{ter}\left[\tanh^{-1}\frac{v}{v_{ter}} - \tanh^{-1}\frac{v_0}{v_{ter}}\right] = gt$$

If the initial velocity is zero, $v_0 = 0$.

$$v_{ter} \tanh^{-1} \frac{v}{v_{ter}} = gt$$

From here,

$$\tanh^{-1} \frac{v}{v_{ter}} = \frac{gt}{v_{ter}}$$

And the speed is obtained as

$$v(t) = v_{ter} \tanh \frac{gt}{v_{ter}}$$

b. By integrating one more time, the equation of motion is obtained,

$$\int_{y_0}^{y} dy' = \int_{0}^{t} \tanh \frac{gt'}{v_{ter}} dt'$$

$$y(t) = y_0 + \frac{(v_{ter})^2}{g} \ln\left[\cosh \frac{gt}{v_{ter}} \right]$$

If the projectile starts with zero velocity and at zero vertical coordinate $y_0 = 0$, then

$$y(t) = \frac{(v_{ter})^2}{g} \ln\left[\cosh \frac{gt}{v_{ter}} \right]$$

PROBLEM 2.7

Imagine a pendulum consisting of a spherical mass m which is placed in front of a large fan as in Figure 2.4. At $t = 0$, the fan begins blowing air at a velocity v_0. Find the equation of motion for the mass considering only quadratic air resistance.

SOLUTION 2.7

The mass experiences the following forces:

$$\vec{F}_{fan} = -cv_o^2 \, \hat{x}$$

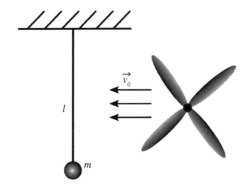

FIGURE 2.4 Pendulum in the presence of a fan.

$$[\vec{F}_{\text{fan}}]_\phi = -cv_0^2 \cos\phi$$

$$\vec{F}_g = -mg\,\hat{y}$$

$$[\vec{F}_g]_\phi = -mg\sin\phi$$

$$\vec{F}_{\text{air}} = -cl^2\dot{\phi}^2\text{sign}(\dot{\phi})\,\hat{\phi}$$

Considering Newton's Second Law

$$\sum F_\phi = ml\ddot{\phi}$$

Using the forces yields

$$-cv_0^2 \cos\phi - mg\sin\phi - c\dot{\phi}^2l^2\text{sign}(\dot{\phi}) = ml\ddot{\phi}$$

Therefore

$$\ddot{\phi} + \frac{cl}{m}\text{sign}(\dot{\phi})\dot{\phi}^2 = -\frac{g}{l}\sin\phi - \frac{cv_0^2}{ml}\cos\phi$$

PROBLEM 2.8

Consider a pendulum of length l displaced by an angle ϕ as shown in Figure 2.5. Find the equations of motion if the mass m experiences

    a. Linear drag.
    b. Quadratic drag.

SOLUTION 2.8

    a. In the case of linear drag

$$\Sigma F_\phi = ma = m\ddot{\phi}l$$

$$-bl\dot{\phi} - mg\sin\phi = m\ddot{\phi}l$$

    Therefore

$$\ddot{\phi} + \frac{b}{m}\dot{\phi} = -\frac{g}{l}\sin\phi$$

    b. In the case of quadratic drag

$$\Sigma F_\phi = ma = m\ddot{\phi}l$$

$$-cl^2\dot{\phi}^2 sign(\dot{\phi}) - mg\sin\phi = m\ddot{\phi}l$$

    Therefore

$$\ddot{\phi} + \frac{cl}{m}\dot{\phi}^2 sign(\dot{\phi}) = -\frac{g}{l}\sin\phi$$

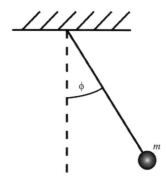

FIGURE 2.5 Simple pendulum.

PROBLEM 2.9

A rock of mass m is thrown downward from a height h at an initial velocity v_0. Find the expression for velocity:

    a. with no air friction
    b. with quadratic air friction cv^2 on the rock

SOLUTION 2.9

    a. According to Newton's Second Law, $\vec{F} = m\ddot{x} = m\dot{v}$. With no air friction, the only force acting on the object is weight. So $\vec{F} = mg = m\dot{v}$. Now solving for v is possible by integrating.

$$\int_{v_0}^{v} dv' = \int_{0}^{t} g\, dt'$$

$$v - v_0 = gt$$

$$v = gt + v_0$$

    b. With air friction, there are two forces acting on the rock: $\vec{F} = mg - cv^2 = m\dot{v}$.

By integrating the equation $\dfrac{m}{c}\dot{v} = \dfrac{mg}{c} - v^2$, an expression for the velocity is obtained.

$$\int_{v_0}^{v} \frac{dv'}{-v'^2 + \dfrac{mg}{c}} = \int_{0}^{t} \frac{c}{m}\, dt'$$

$$\sqrt{\frac{c}{mg}}\left(\tanh^{-1}\left(v\sqrt{\frac{c}{mg}} \right) - \tanh^{-1}\left(v_0 \sqrt{\frac{c}{mg}} \right) \right) = \frac{c}{m}t$$

PROBLEM 2.10

A block of mass m is launched with initial velocity v_0 down an incline at an angle θ with the horizontal. The block starts from rest and is subject to linear drag $\vec{f} = b\vec{v}$. The friction coefficient between the incline and the block is μ. Find an expression for the velocity of the block.

SOLUTION 2.10

From Newton's Second Law, $\vec{F_y} = m\ddot{y} = 0$ and $\vec{F_x} = m\ddot{x} = m\dot{v}_x$. The forces acting on the block (Figure 2.6) are $\vec{F_y} = N - mg\cos\theta$ and $\vec{F_x} = -bv_x + mg\sin\theta - \mu N$. First, a value for N is found.

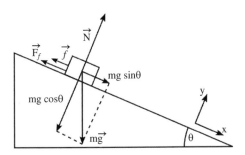

FIGURE 2.6 Block sliding down an incline, with drag and friction forces.

$$N - mg\cos\theta = 0$$

$$N = mg\cos\theta$$

An expression for the velocity of the block is obtained by integrating $\dot{v}_x = -\dfrac{b}{m}v_x + g\sin\theta - \mu g\cos\theta$.

$$\int_{v_0}^{v_x} \frac{dv'_x}{-\dfrac{b}{m}v'_x + g\sin\theta - \mu g\cos\theta} = \int_0^t dt'$$

$$-\frac{m}{b}\left(\ln\left|-\frac{b}{m}v_x + g\sin\theta - \mu g\cos\theta\right| - \ln\left|-\frac{b}{m}v_0 + g\sin\theta - \mu g\cos\theta\right|\right) = t$$

$$\ln\frac{\left|-\dfrac{b}{m}v_x + g\sin\theta - \mu g\cos\theta\right|}{\left|-\dfrac{b}{m}v_0 + g\sin\theta - \mu g\cos\theta\right|} = -\frac{b}{m}t$$

$$\frac{\left|-\dfrac{b}{m}v_x + g\sin\theta - \mu g\cos\theta\right|}{\left|-\dfrac{b}{m}v_0 + g\sin\theta - \mu g\cos\theta\right|} = e^{-\frac{b}{m}t}$$

$$\left|-\frac{b}{m}v_x + g\sin\theta - \mu g\cos\theta\right| = \left|-\frac{b}{m}v_0 + g\sin\theta - \mu g\cos\theta\right|e^{-\frac{b}{m}t}$$

If $-\dfrac{b}{m}v_x + g\sin\theta - \mu g\cos\theta > 0$

$$-\frac{b}{m}v_x = -g\sin\theta + \mu g\cos\theta + \left| -\frac{b}{m}v_0 + g\sin\theta - \mu g\cos\theta \right| e^{-\frac{b}{m}t}$$

$$v_x(t) = \frac{m}{b}\left(g\sin\theta - \mu g\cos\theta - \left| -\frac{b}{m}v_0 + g\sin\theta - \mu g\cos\theta \right| e^{-\frac{b}{m}t} \right)$$

If $-\dfrac{b}{m}v_x + g\sin\theta - \mu g\cos\theta < 0$

$$\frac{b}{m}v_x = g\sin\theta - \mu g\cos\theta + \left| -\frac{b}{m}v_0 + g\sin\theta - \mu g\cos\theta \right| e^{-\frac{b}{m}t}$$

$$v_x(t) = \frac{m}{b}\left(g\sin\theta - \mu g\cos\theta + \left| -\frac{b}{m}v_0 + g\sin\theta - \mu g\cos\theta \right| e^{-\frac{b}{m}t} \right)$$

PROBLEM 2.11

A cart is moving on a horizontal surface with initial velocity v_0 subject to a drag force of the type $f(v) = -cv^{\frac{3}{2}}$. Find the speed of the cart as a function of time.

SOLUTION 2.11

The net force is the drag force.

$$\vec{F}_{net} = m\vec{g} + \vec{N} + \vec{f}_{drag} = \vec{f}_{drag} = -cv^{\frac{3}{2}}\,\hat{v}$$

$$m\dot{v} = -cv^{\frac{3}{2}}$$

$$m\frac{dv}{dt} = -cv^{\frac{3}{2}}$$

$$\frac{dv}{v^{\frac{3}{2}}} = -\frac{c}{m}dt$$

The velocity is obtained as before, by integrating on both sides the previous equation over velocity, from v_0 to v, and over time, from 0 to t.

$$\int_{v_0}^{v} \frac{dv'}{(v')^{\frac{3}{2}}} = -\frac{c}{m}\int_0^t dt'$$

$$-\frac{2}{\sqrt{v'}}\Big|_{v_0}^{v} = -\frac{c}{m}t$$

$$\frac{1}{\sqrt{v}} - \frac{1}{\sqrt{v_0}} = \frac{ct}{2m}$$

$$v(t) = \frac{1}{\left(\dfrac{1}{\sqrt{v_0}} + \dfrac{ct}{2m}\right)^2}$$

$$v(t) = \frac{v_0}{\left(1 + \dfrac{ct\sqrt{v_0}}{2m}\right)^2}$$

As expected, when time approaches infinity, the speed goes to zero.

PROBLEM 2.12

An object of mass m is coasting horizontally with initial horizontal speed v_0 under the influence of both linear and quadratic air drag. Obtain the speed as a function of time for this system.

SOLUTION 2.12

The object is cruising under the influence of both linear and quadratic air drag, which means that both drag forces have the same order or magnitude and neither of them can be neglected. Here, Newton's Second Law is

$$m\dot{v} = -bv - cv^2$$

$$m\frac{dv}{dt} = -bv - cv^2$$

$$m\frac{dv}{bv + cv^2} = -dt$$

$$m\frac{dv}{cv\left(\dfrac{b}{c} + v\right)} = -dt$$

Using the integral (with $a \neq b$),

$$\int \frac{1}{(x+a)(x+b)}\,dx = \frac{1}{b-a}[\ln(a+x) - \ln(b+x)]$$

$$\frac{m}{c}\int_{v_0}^{v}\frac{dv'}{v'\left(v'+\dfrac{b}{c}\right)} = -\int_{0}^{t}dt'$$

$$\frac{m}{c}\frac{1}{\left(\dfrac{b}{c}\right)}\left[\ln v' - \ln\left(\frac{b}{c}+v'\right)\right]\Bigg|_{v_0}^{v} = -t$$

$$\frac{m}{b}\ln\frac{v'}{\left(\dfrac{b}{c}+v'\right)}\Bigg|_{v_0}^{v} = -t$$

$$\frac{m}{b}\left[\ln\frac{v}{\left(\dfrac{b}{c}+v\right)} - \ln\frac{v_0}{\left(\dfrac{b}{c}+v_0\right)}\right] = -t$$

By using the properties of the logarithms, $\ln A - \ln B = \ln\dfrac{A}{B}$, it follows that

$$\ln\frac{\dfrac{v}{\left(\dfrac{b}{c}+v\right)}}{\dfrac{v_0}{\left(\dfrac{b}{c}+v_0\right)}} = -\frac{b}{m}t$$

$$\ln\frac{\dfrac{b}{cv}+1}{\dfrac{b}{cv_0}+1} = \frac{b}{m}t$$

$$\frac{b}{cv}+1 = \left(\frac{b}{cv_0}+1\right)e^{\left(\frac{b}{m}t\right)}$$

With some algebra rearrangement,

$$v(t) = \frac{b}{c}\left[-1+\left(\frac{b}{cv_0}+1\right)e^{\left(\frac{b}{m}t\right)}\right]^{-1}$$

$$v(t) = \frac{b}{c} \frac{1}{-1 + \left(\dfrac{b}{cv_0} + 1\right) e^{\left(\frac{b}{m}t\right)}}$$

With a little algebra rearrangement, the speed becomes

$$v(t) = \frac{b}{c} \frac{1}{-1 + \left(\dfrac{b}{cv_0} + 1\right) \dfrac{1}{e^{-\left(\frac{b}{m}t\right)}}}$$

$$v(t) = \frac{b}{c} \frac{e^{-\left(\frac{b}{m}t\right)}}{-e^{-\left(\frac{b}{m}t\right)} + \left(\dfrac{b}{cv_0} + 1\right)}$$

$$v(t) = \frac{b}{c} \frac{e^{-\left(\frac{b}{m}t\right)}}{1 + \dfrac{b}{cv_0} - e^{-\left(\frac{b}{m}t\right)}}$$

By checking the result for the validity at initial time zero, we obtain the initial velocity $(t = 0) = v_0$, as expected.

When the time approaches infinity, the velocity is closer to zero, therefore the quadratic air drag becomes negligible compared to the linear drag, which will be the main contribution for velocity close to zero. The speed is exponentially decreasing, similarly to problems in which only the linear air drag is present.

PROBLEM 2.13

Suppose there exists a nonstandard sphere which experiences quadratic drag in one orientation and linear drag in an orthogonal one. Given mass m and initial velocity v_0, it is launched at an angle θ. Determine its velocity as a function of time and angle for the orientation of linear drag in the x direction, quadratic drag in the y. Take $f_{\text{lin}}(v) = bv$ and $f_{\text{quad}} = cv^2$ for constants b, c with the appropriate units.

SOLUTION 2.13

Consider the x and y directions separately. In the x direction

$$\sum F_x = m\dot{v}_x$$

$$-bv_x = m\frac{dv_x}{dt}$$

$$\frac{dv_x}{dt} = -\frac{b}{m}v_x$$

so for a constant A

$$v_x(t) = A\, e^{-\frac{b}{m}t}$$

Since $v_x(0) = v_0\cos(\theta)$,

$$A = v_0\cos(\theta)$$

Therefore

$$v_x(t) = v_0\cos(\theta)e^{-\frac{b}{m}t}$$

Now consider the y direction. This must be split into two parts, the ascent motion and the descent motion. During the ascent, gravity and the drag will be in the same direction, where during the descent, they are in opposite direction. Considering the ascent first

$$\sum F_y = m\dot{v}_y$$

$$-mg - cv^2 = m\frac{dv_y}{dt}$$

$$\frac{dv_y}{dt} = -\left(g + \frac{c}{m}v^2\right) = -\frac{c}{m}\left(\frac{gm}{c} + v^2\right)$$

$$\frac{dv_y}{\frac{gm}{c} + v^2} = -\frac{c}{m}dt$$

$$\sqrt{\frac{c}{gm}}\tan^{-1}\left(\sqrt{\frac{c}{gm}}v_y\right) = -\frac{c}{m}t + A$$

where A is a constant. Considering $v_y(0) = v_0\sin(\theta)$

$$A = \sqrt{\frac{c}{gm}}\tan^{-1}\left(\sqrt{\frac{c}{gm}}v_0\sin(\theta)\right)$$

Now solving for v_y

$$\sqrt{\frac{c}{gm}}\tan^{-1}\left(\sqrt{\frac{c}{gm}}v_y\right) = -\frac{c}{m}t + \sqrt{\frac{c}{gm}}\tan^{-1}\left(\sqrt{\frac{c}{gm}}v_0\sin(\theta)\right)$$

$$\tan^{-1}\left(\sqrt{\frac{c}{gm}}v_y\right) = -\sqrt{\frac{gc}{m}}t + \tan^{-1}\left(\sqrt{\frac{c}{gm}}v_0\sin(\theta)\right)$$

$$\sqrt{\frac{c}{gm}}v_y = \tan\left(-\sqrt{\frac{gc}{m}}t + \tan^{-1}\left(\sqrt{\frac{c}{gm}}v_0\sin(\theta)\right)\right)$$

$$v_y = \sqrt{\frac{gm}{c}}\left(\frac{\sqrt{\frac{c}{gm}}v_0\sin(\theta) - \tan\left(\sqrt{\frac{gc}{m}}t\right)}{1 + \sqrt{\frac{c}{gm}}v_0\sin(\theta)\tan\left(\sqrt{\frac{gc}{m}}t\right)}\right)$$

On the descent, the drag force is in the opposite direction. Utilizing the setup of the ascent,

$$\frac{dv_y}{dt} = -\left(g - \frac{c}{m}v^2\right) = -\frac{c}{m}\left(\frac{gm}{c} - v^2\right)$$

$$\frac{dv_y}{\frac{gm}{c} - v^2} = -\frac{c}{m}dt$$

$$\frac{1}{2}\sqrt{\frac{c}{gm}}\ln\left|\frac{\sqrt{\frac{gm}{c}} + v_y}{\sqrt{\frac{gm}{c}} - v_y}\right| = -\frac{c}{m}t + A$$

$$\ln\left|\frac{\sqrt{\frac{gm}{c}} + v_y}{\sqrt{\frac{gm}{c}} - v_y}\right| = -2\sqrt{\frac{gc}{m}}t + A$$

$$\left|\frac{\sqrt{\frac{gm}{c}} + v_y}{\sqrt{\frac{gm}{c}} - v_y}\right| = Ae^{-2\sqrt{\frac{gc}{m}}t}$$

Since the ball is descending, the initial velocity is zero, so at time $t = 0$

$$\left| \frac{\sqrt{\frac{gm}{c}}}{\sqrt{\frac{gm}{c}}} \right| = A$$

$$A = 1$$

Also $v_y \leq 0$ so

$$\left| \frac{\sqrt{\frac{gm}{c}} + v_y}{\sqrt{\frac{gm}{c}} - v_y} \right| = e^{-2\sqrt{\frac{gc}{m}}t}$$

Since terminal velocity is $\sqrt{\frac{gm}{c}}$, $|v_y| \leq \sqrt{\frac{gm}{c}}$,

$$\frac{\sqrt{\frac{gm}{c}} + v_y}{\sqrt{\frac{gm}{c}} - v_y} = e^{-2\sqrt{\frac{gc}{m}}t}$$

$$\sqrt{\frac{gm}{c}} + v_y = \sqrt{\frac{gm}{c}}\, e^{-2\sqrt{\frac{gc}{m}}t} - v_y e^{-2\sqrt{\frac{gc}{m}}t}$$

$$v_y\left(1 + e^{-2\sqrt{\frac{gc}{m}}t}\right) = \sqrt{\frac{gm}{c}}\left(e^{-2\sqrt{\frac{gc}{m}}t} - 1\right)$$

Therefore

$$v_y(t) = \sqrt{\frac{gm}{c}}\left(\frac{e^{-2\sqrt{\frac{gc}{m}}t} - 1}{e^{-2\sqrt{\frac{gc}{m}}t} + 1}\right) = \sqrt{\frac{gm}{c}}\left(\frac{e^{-\sqrt{\frac{gc}{m}}t} - e^{\sqrt{\frac{gc}{m}}t}}{e^{-\sqrt{\frac{gc}{m}}t} + e^{\sqrt{\frac{gc}{m}}t}}\right) = -\sqrt{\frac{gm}{c}}\tanh\sqrt{\frac{gc}{m}}t$$

PROBLEM 2.14

Consider a sailboat with a sail which experiences quadratic drag on one side and linear drag on the other. On its maiden voyage, there is a curious wind blowing at a velocity $v(t) = v_0 \sin(\omega t)$. Determine the velocity of the boat as a function of time if the wind hits the side with the quadratic drag. Assume no friction due to the water.

SOLUTION 2.14

The wind is pushing with a force

$$\vec{F}_{applied} = cv_0^2 \sin^2(\omega t)\,\hat{x}$$

And the air resistance is given by

$$\vec{F}_{air} = -bv\,\hat{x}$$

Therefore,

$$\Sigma F_x = m\dot{v}$$

$$cv_0^2 \sin^2(\omega t) - bv = m\dot{v}$$

Since the mass is constant, introduce $c' = \dfrac{c}{m}$.

$$\dot{v} + \frac{b}{m}v = c'v_0^2 \sin^2(\omega t)$$

This can be solved using the method of integrating factors. Consider the integrating factor

$$\mu = e^{\int \frac{b}{m}\,dt} = e^{\frac{b}{m}t}$$

Now,

$$\frac{d}{dt}[\mu v] = \mu\, c'v_0^2 \sin^2(\omega t)$$

$$v = \frac{c'v_0^2}{\mu}\int \mu \sin^2(\omega t)\, dt$$

$$v = c'v_0^2 e^{-\frac{b}{m}t}\int e^{\frac{b}{m}t} \sin^2(\omega t)\, dt$$

This can be solved using integration by parts. It will be helpful to express $\sin^2(\omega t)$ differently using the following equations:

$$\cos^2 x - \sin^2 x = \cos(2x)$$

$$\cos^2 x = 1 - \sin^2 x$$

so

$$\sin^2(\omega t) = \frac{1}{2}(1 - \cos(2\omega t))$$

Therefore, the integral to solve is

$$v = \frac{1}{2}c'v_0^2 e^{-\frac{b}{m}t} \int e^{\frac{b}{m}t} - e^{\frac{b}{m}t}\cos(2\omega t)\,dt$$

$$v = \frac{1}{2}c'v_0^2 e^{-\frac{b}{m}t}\frac{m}{b}e^{\frac{b}{m}t} - \frac{1}{2}c'v_0^2 e^{-\frac{b}{m}t} \int e^{\frac{b}{m}t}\cos(2\omega t)\,dt$$

$$v = \frac{c'v_0^2 m}{2b} - \frac{1}{2}c'v_0^2 e^{-\frac{b}{m}t} \int e^{\frac{b}{m}t}\cos(2\omega t)\,dt$$

Now to solve for the integral $\int e^{\frac{b}{m}t}\cos(2\omega t)\,dt$ consider the following for integration by parts

$$u = \cos(2\omega t),\ du = -2\omega\sin(2\omega t)dt,\ v = \frac{m}{b}e^{\frac{b}{m}t},\ dv = e^{\frac{b}{m}t}\,dt$$

so

$$\int e^{\frac{b}{m}t}\cos(2\omega t)\,dt = \frac{m}{b}e^{\frac{b}{m}t}\cos(2\omega t) + \frac{2\omega m}{b}\int e^{\frac{b}{m}t}\sin(2\omega t)dt$$

Again, integration by parts will be used to solve $\int e^{\frac{b}{m}t}\sin(2\omega t)\,dt$.

$$u = \sin(2\omega t),\ du = 2\omega\cos(2\omega t)dt,\ v = \frac{m}{b}e^{\frac{b}{m}t},\ dv = e^{\frac{b}{m}t}\,dt$$

so

$$\int e^{\frac{b}{m}t}\sin(2\omega t)dt = \frac{m}{b}e^{\frac{b}{m}t}\sin(2\omega t) + \frac{2\omega m}{b}\int e^{\frac{b}{m}t}\cos(2\omega t)dt$$

Combining the two integrations by parts yields

$$\int e^{\frac{b}{m}t}\cos(2\omega t)\,dt = \frac{m}{b}e^{\frac{b}{m}t}\cos(2\omega t) + \frac{2\omega m}{b}\left(\frac{m}{b}e^{\frac{b}{m}t}\sin(2\omega t) + \frac{2\omega m}{b}\int e^{\frac{b}{m}t}\cos(2\omega t)\,dt\right)$$

$$\int e^{\frac{b}{m}t}\cos(2\omega t)\,dt = \frac{m}{b}e^{\frac{b}{m}t}\cos(2\omega t) + \frac{2\omega m^2}{b^2}e^{\frac{b}{m}t}\sin(2\omega t) + \frac{4\omega^2 m^2}{b^2}\int e^{\frac{b}{m}t}\cos(2\omega t)\,dt$$

$$\left(1+\frac{4\omega^2 m^2}{b^2}\right)\int e^{\frac{b}{m}t}\cos(2\omega t)\,dt = \frac{m}{b}e^{\frac{b}{m}t}\cos(2\omega t) + \frac{2\omega m^2}{b^2}e^{\frac{b}{m}t}\sin(2\omega t)$$

$$\int e^{\frac{b}{m}t}\cos(2\omega t)\,dt = \left(\frac{b^2}{b^2+4\omega^2 m^2}\right)\left(\frac{m}{b}e^{\frac{b}{m}t}\cos(2\omega t) + \frac{2\omega m^2}{b^2}e^{\frac{b}{m}t}\sin(2\omega t)\right)$$

$$\int e^{\frac{b}{m}t}\cos(2\omega t)\,dt = \left(\frac{mb}{b^2+4\omega^2 m^2}\right)e^{\frac{b}{m}t}\left(\cos(2\omega t) + \frac{2\omega m}{b}\sin(2\omega t)\right)$$

An expression for v is now given by

$$v = \frac{c'v_0^2 m}{2b} - \frac{1}{2}c'v_0^2 e^{-\frac{b}{m}t}\left(\frac{mb}{b^2+4\omega^2 m^2}\right)e^{\frac{b}{m}t}\left(\cos(2\omega t) + \frac{2\omega m}{b}\sin(2\omega t)\right)$$

$$v = \frac{c'v_0^2 m}{2}\left(\frac{1}{b} - \frac{b}{b^2+4\omega^2 m^2}\left(\cos(2\omega t) + \frac{2\omega m}{b}\sin(2\omega t)\right)\right)$$

3 Momentum and Angular Momentum

3.1 THEORY

This chapter introduces linear momentum, moment of inertia, center of mass, angular momentum, and the principles of conservation of linear momentum and of angular momentum.

3.1.1 LINEAR MOMENTUM

The variation of the total linear momentum $\dfrac{d\vec{P}}{dt} = \dot{\vec{P}}$ is equal to the external force $\vec{F}_{ext}$ acting on the system:

$$\dot{\vec{P}} = \vec{F}_{ext}$$

with $\vec{P} = \vec{p}_1 + \vec{p}_2 + \ldots + \vec{p}_N = \sum_{i=0}^{N} \vec{p}_i$

Conservation of the linear momentum: if the net external force applied on the system is zero, the total linear momentum of the system is conserved.

3.1.2 ROCKETS

A rocket of initial mass m_0 and initial velocity v_0 (on positive x direction), with a fuel with exhaust velocity v_{ex} (on negative x direction) with respect to the rocket and without external forces, acquires a velocity v at the moment the mass becomes m

$$v = v_0 + v_{ex} \ln \frac{m_0}{m}$$

The thrust of the rocket is defined as the product of the rate at which the mass is ejected, $-\dot{m}$ with the speed of the exhaust v_{ex}, thrust $= -\dot{m} v_{ex}$, where it is important to note that the mass becomes smaller with the ejection of the fuel, and $\dot{m} < 0$.

3.1.3 CENTER OF MASS

For a system of N discrete particles of mass m_i and position vector $\vec{r}_i$, the center of the mass is

$$\vec{R} = \frac{1}{M} \sum_{i=0}^{N} m_i \vec{r}_i = \frac{m_1 \vec{r}_1 + m_2 \vec{r}_2 + \ldots + m_N \vec{r}_N}{M}$$

DOI: 10.1201/9781003365709-3

where M is the total mass of the system,

$$M = \sum_{i=0}^{N} m_i$$

For a continuous system of mass M and density ρ,

$$\vec{R} = \frac{1}{M} \int \vec{r} \, dm = \frac{1}{M} \int \rho \vec{r} \, dV$$

3.1.4 MOMENT OF INERTIA

The moment of inertia for a discrete system is obtained as

$$I = \sum_{\alpha=1}^{N} m_\alpha s_\alpha^2$$

with s_α the distances of the mass m_α from the axis of rotation.

For a continuous system, the sum is transformed into an integral, with ρ the density of the object, s the distance from the axis of rotation, and dV the element of volume.

$$I = \int \rho \, s^2 dV$$

Angular momentum is given by

$$\vec{l} = \vec{r} \times \vec{p}$$

The rate of the angular momentum $\dfrac{d\vec{l}}{dt} = \dot{\vec{l}}$

$$\dot{\vec{l}} = \vec{r} \times \vec{F} = \vec{\Gamma}$$

where $\vec{\Gamma}$ is the net applied torque about the origin.

3.1.5 PRINCIPLE OF CONSERVATION OF ANGULAR MOMENTUM

If the net external torque $\vec{\Gamma}$ acting on a particle is zero, then the total angular momentum of the particle is constant, $\vec{l} = \text{constant}$.

Angular momentum for a system of N particles $\vec{L}$ is

$$\vec{L} = \sum_{i=0}^{N} \vec{l}_i = \sum_{i=0}^{N} \vec{r}_i \times \vec{p}_i$$

with

$$\dot{\vec{L}} = \sum_{i=0}^{N} \dot{\vec{l}}_i = \sum_{i=0}^{N} \vec{r}_i \times \vec{F}_i = \vec{\Gamma}^{\text{ext}}$$

3.1.6 PRINCIPLE OF CONSERVATION OF THE ANGULAR MOMENTUM FOR A SYSTEM OF N PARTICLES

If the net external torque $\vec{\Gamma}^{\text{ext}}$ acting on a system of N particle is zero, then the total angular momentum of the particle is constant, $\vec{L} = \text{constant}$.

3.2 PROBLEMS AND SOLUTIONS

PROBLEM 3.1

A beginner bodybuilder accidentally put a mass of 20 kg at one end of his bar and a mass of 30 kg at the other end. The bar is 1.5 m long. Find the center of mass of the system.

SOLUTION 3.1

This problem consists of finding the center of mass of a two-particle system (Figure 3.1).

$$X = \frac{1}{M} \sum_{\alpha=1}^{2} m_\alpha x_\alpha$$

$$X = \frac{1}{20\,\text{kg} + 30\,\text{kg}} (20\,\text{kg} \cdot 0\,\text{m} + 30\,\text{kg} \cdot 1.5\,\text{m})$$

$$X = 0.9\,\text{m}$$

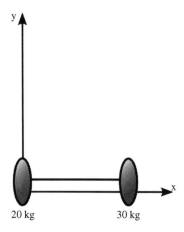

FIGURE 3.1 Two masses of 20 kg and 30 kg, respectively, at the ends of a bar.

PROBLEM 3.2

Two blocks of mass m_1 and m_2, with initial velocity v_1 and v_2, respectively, meet in a perfectly inelastic collision. Then, a third block of mass m_3 and velocity v_3 hits the two blocks in a perfectly inelastic collision. Using conservation of momentum, find the velocity of the three blocks, assuming no external forces (one dimensional case).

SOLUTION 3.2

The collision of m_1 and m_2 is first considered. Because of the conservation of momentum, $\vec{p}_{\text{initial}} = \vec{p}_{\text{final}}$. Thus,

$$m_1 v_1 + m_2 v_2 = (m_1 + m_2) v_{(12)}$$

The velocity of the two blocks is

$$v_{(12)} = \frac{m_1 v_1 + m_2 v_2}{m_1 + m_2}$$

Now, the collision with the third block is studied. Again, conservation of momentum gives $\vec{p}_{\text{initial}} = \vec{p}_{\text{final}}$

$$(m_1 + m_2) v_{(12)} + m_3 v_3 = (m_1 + m_2 + m_3) v$$

The velocity of the three blocks is

$$v = \frac{m_1 v_1 + m_2 v_2 + m_3 v_3}{m_1 + m_2 + m_3}$$

PROBLEM 3.3

Consider two particles of mass m_1 and m_2. Find the amount of kinetic energy loss during an inelastic collision if the particles are initially traveling at $\vec{v}_1$ and $\vec{v}_2$. Considering the case when $m_1 = m_2 = m$, can this energy loss be ignored in any situation?

SOLUTION 3.3

The initial kinetic energy is given by

$$KE_i = \frac{1}{2}(m_1 v_1^2 + m_2 v_2^2)$$

and the final kinetic energy is given by

$$KE_f = \frac{1}{2}(m_1 + m_2)v_f^2$$

where v_f is the final energy of the system. An expression for v_f can be found by considering conservation of momentum:

$$\vec{p}_i = \vec{p}_f$$

$$m_1 \vec{v}_1 + m_2 \vec{v}_2 = (m_1 + m_2)\vec{v}_f$$

with

$$\vec{v}_f = \frac{m_1 \vec{v}_1 + m_2 \vec{v}_2}{(m_1 + m_2)}$$

Therefore,

$$\vec{v}_f^2 = \frac{1}{(m_1 + m_2)^2}((m_1 \vec{v}_1 + m_2 \vec{v}_2) \cdot (m_1 \vec{v}_1 + m_2 \vec{v}_2))$$

$$= \frac{1}{(m_1 + m_2)^2}(m_1^2 v_1^2 + m_2^2 v_2^2 + 2m_1 m_2 \vec{v}_1 \cdot \vec{v}_2)$$

The final kinetic energy is now given by

$$KE_f = \frac{1}{2}\frac{1}{(m_1 + m_2)}(m_1^2 v_1^2 + m_2^2 v_2^2 + 2m_1 m_2 \vec{v}_1 \cdot \vec{v}_2)$$

and the energy lost is therefore

$$KE_i - KE_f = \frac{1}{2}\left(m_1 v_1^2 + m_2 v_2^2 - \frac{m_1^2 v_1^2 + m_2^2 v_2^2 + 2m_1 m_2 \, \vec{v}_1 \cdot \vec{v}_2}{m_1 + m_2} \right)$$

$$= \frac{1}{2}\left(\frac{m_1 m_2 \left(v_1^2 + v_2^2 - 2\vec{v}_1 \cdot \vec{v}_2 \right)}{m_1 + m_2} \right)$$

Considering the case where $m_1 = m_2 = m$, the energy lost becomes

$$\delta KE = KE_i - KE_f = \frac{1}{2} m(v_1^2 + v_2^2 - 2\vec{v}_1 \cdot \vec{v}_2) = \frac{1}{2} m(\vec{v}_1 - \vec{v}_2)^2$$

Considering $\vec{v}_1 = \vec{v}_2 + \vec{\delta}_v$,

$$\delta KE = \frac{1}{2} m \delta_v^2$$

which is negligible provided $|\delta_v^2| \ll 1$. This amounts to the boring scenario of two masses traveling in approximately the same direction at approximately the same speed and simply sitting together. This shows that treating inelastic collisions via energy consideration may not be the best idea as the energy lost is not negligible.

PROBLEM 3.4

Analyze the motion of a rocket starting with initial mass m_0, which accelerated from rest. Obtain the momentum versus mass, $p(m)$ and find the mass for which you obtain the maximum momentum. What was the mass of the fuel consumed? Find the maximum momentum.

SOLUTION 3.4

The speed of the single stage rocket is dependent on the speed of the exhaust with respect to the rocket v_{ex}, initial mass of the rocket and fuel m_0, and the final mass of the rocket m

$$v = v_{ex} \ln \frac{m_0}{m}$$

And the linear momentum

$$p(m) = mv = mv_{ex} \ln \frac{m_0}{m}$$

The momentum is maximum when the derivative with respect to the mass is zero.

$$\frac{dp(m)}{dm} = v_{ex} \ln \frac{m_0}{m} + m v_{ex} \frac{m}{m_0}\left(-\frac{m_0}{m^2}\right) = v_{ex}\left(\ln \frac{m_0}{m} - 1\right)$$

The derivative of momentum with respect to mass is zero when

$$v_{ex}\left(\ln \frac{m_0}{m} - 1\right) = 0$$

which implies

$$\ln \frac{m_0}{m} = 1$$

therefore, the mass of the rocket is

$$m = \frac{m_0}{e}$$

where e is the base of the natural logarithm.
The mass of the fuel consumed is

$$m_{fuel} = m_0 - m = m_0 - \frac{m_0}{e} = m_0\left(1 - \frac{1}{e}\right) \approx 0.63\, m_0$$

Therefore, for a rocket to reach the highest momentum, 63% of the rocket's initial mass should be consumed as fuel. In general, the rocket's velocities are improved by a multiple stage system, when the fuel tank is discarded, as discussed in another rocket problem.
The momentum corresponding to the maximum point is

$$p_{max} = p\left(\frac{m_0}{e}\right) = \frac{m_0}{e} v\left(\frac{m_0}{e}\right) = \frac{m_0}{e} v_{ex} \ln \frac{m_0}{\dfrac{m_0}{e}} = \frac{m_0 v_{ex}}{e}$$

PROBLEM 3.5
Find the center of mass of

a. A cone of radius R and height h, with its base on the xy plane.
b. A hemisphere of radius R, with its base on the xy plane.

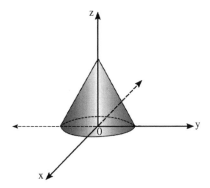

FIGURE 3.2   Cone with its base on the xy plane, centered at origin.

SOLUTION 3.5

a. Because of the cone's symmetry (Figure 3.2), $X = 0$ and $Y = 0$.

$$Z = \frac{1}{M} \int \rho z \, dV$$

The integration is performed using cylindrical coordinates, from 0 to $r = \frac{h-z}{h} R$ over r, from 0 to 2π over θ, and from 0 to h over z.

$$Z = \frac{\rho}{M} \int_0^h \int_0^{2\pi} \int_0^{\frac{h-z}{h}R} z r \, dr d\theta dz$$

$$Z = \frac{2\pi\rho}{M} \int_0^h \frac{R^2}{2} \left(\frac{h-z}{h} \right)^2 dz$$

$$Z = \frac{\pi\rho R^2}{Mh^2} \int_0^h (zh^2 - 2hz^2 + z^3) dz$$

$$Z = \frac{\pi\rho R^2 h^2}{12M}$$

Since $\rho = M/V$,

$$Z = \frac{\pi R^2 h^2}{12V}$$

The volume of a cone is $V = \pi R^2 \dfrac{h}{3}$.

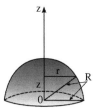

FIGURE 3.3 Half sphere with its base on the *xy* plane.

$$Z = \frac{\pi R^2 h^2}{12\pi R^2 h/3}$$

$$Z = \frac{h}{4}$$

b. Because of the hemisphere's symmetry (Figure 3.3), $X = 0$ and $Y = 0$.

$$Z = \frac{1}{M}\int \rho z \, dV$$

The integration is performed using cylindrical coordinates, from 0 to $r = \sqrt{R^2 - z^2}$ over r, from 0 to 2π over θ, and from 0 to R over z.

$$Z = \frac{\rho}{M}\int_0^R \int_0^{2\pi} \int_0^{\sqrt{R^2-z^2}} zr \, dr d\theta dz$$

$$Z = \frac{2\pi\rho}{M}\int_0^h z\frac{R^2 - z^2}{2} dz$$

$$Z = \frac{\pi R^4}{4V}$$

The volume of a hemisphere is $V = \frac{1}{2}\left(\frac{4}{3}\pi R^3\right)$.

$$Z = \frac{\pi R^4}{4\left(\frac{4}{3}\pi R^3\right)\frac{1}{2}}$$

$$Z = \frac{3R}{8}$$

PROBLEM 3.6

Find the center of mass of a composite solid made of a cone of height h and radius R and a hemisphere of same radius, joined at their base.

SOLUTION 3.6

The cone and the hemisphere (Figure 3.4) can be considered as particles, using their respective center of mass. The center of mass of the cone is $Z_c = \dfrac{h}{4}$ and the one of

the hemisphere is $Z_h = \dfrac{3R}{8}$, as shown in Problem 3.5.

The center of mass of the composite solid is found by computing

$$Z = \frac{1}{\displaystyle\sum_{\alpha=1}^{2} m_\alpha} \sum_{\alpha=1}^{2} m_\alpha z_\alpha .$$

$$Z = \frac{m_c Z_c + m_h Z_h}{m_c + m_h}$$

$$Z = \frac{\left(\rho \pi R^2 \dfrac{h}{3}\right)\left(\dfrac{h}{4}\right) + \left(\dfrac{2}{3}\rho \pi R^3\right)\left(\dfrac{-3}{8}R\right)}{\rho \pi R^2 \dfrac{h}{3} + \dfrac{2}{3}\rho \pi R^3}$$

$$Z = \frac{\dfrac{h^2}{12} - \dfrac{R^2}{4}}{\dfrac{h}{3} + \dfrac{2R}{3}}$$

$$Z = \frac{h^2 - 3R^2}{4(h + 2R)}$$

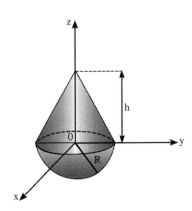

FIGURE 3.4   Cone and half sphere connected to each other from their base.

PROBLEM 3.7

a. A one-stage rocket consumes 70% of the initial mass as fuel. Find the final speed of the rocket considering that it starts from rest and that the exhaust has the speed v_{ex} with respect to the rocket.

b. A two-stage rocket starting from rest consumes 25% of the initial mass as fuel in the first stage. At the end of the first stage, the tank with a mass of 20% of the initial mass is ejected. In the second stage, the rocket consumes another 25% of the initial mass as fuel. Find the speed of the rocket at the end of the stage I and at the end of the stage II.

SOLUTION 3.7

a. The final speed of the one-stage rocket is

$$v = v_{ex} \ln \frac{m_0}{m}$$

where v is the final speed of the rocket, m_0 is the initial mass of the rocket plus fuel, and m is the final mass of the rocket. Since a mass of 0.7 m_0 is consumed as fuel, the final mass of the rocket is 0.3 m_0

$$v = v_{ex} \ln \frac{m_0}{m} = v_{ex} \ln \frac{m_0}{m_0 - 0.7 m_0} = v_{ex} \ln \frac{1}{0.3} = v_{ex} \ln 3.33$$

b. During the first stage, the rocket will consume fuel with the mass of 0.25 m_0 and will gain speed v_1

$$v_1 = v_{ex} \ln \frac{m_0}{m} = v_{ex} \ln \frac{m_0}{m_0 - 0.25 m_0} = v_{ex} \ln \frac{1}{0.75} = v_{ex} \ln \frac{4}{3}$$

After the first stage the rocket ejects the tank, having an even smaller mass $m_0(1 - 0.25 - 0.2) = 0.55 m_0$, and starts this stage with the initial velocity v_1, at the end of the stage is gaining the velocity v_2

$$v_2 = v_1 + v_{ex} \ln \frac{0.55 m_0}{0.3 m_0} = v_{ex} \ln \frac{4}{3} + v_{ex} \ln \frac{0.55}{0.3} = v_{ex} \ln \left(\frac{4}{3} + \frac{0.55}{0.3} \right) = 0.89 v_{ex}$$

The property of logarithms is used here: $\ln A + \ln B = \ln AB$.

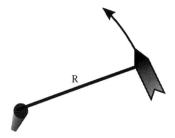

FIGURE 3.5   A rocket connected to a pole.

PROBLEM 3.8

A rocket is tethered to a pole such that it moves in a circular orbit (Figure 3.5). At $t = 0$, the rocket begins burning fuel in a way that decreases its mass in accordance with the equation $m(t) = m_0 e^{-t\lambda}$.

If the initial rope length is R, initial mass is m_0, and initial velocity is v, find an expression for the rope length, as a function of time, that keeps the angular momentum constant. Assume the rocket thrust is such that the rocket's velocity increases as $v(t) = v + \alpha t$.

SOLUTION 3.8

The initial angular momentum is given by

$$\vec{l} = \vec{r} \times \vec{p}$$

Since $\vec{r} \perp \vec{p}$

$$\left|\vec{l}\right| = l = R m_0 v$$

At $t > 0$, the angular momentum becomes

$$l_{t>0} = r(t) m_0 e^{-\lambda t}(v + \alpha t)$$

keeping $l_{t>0} = l$ requires

$$r(t) m_0 e^{-\lambda t}(v + \alpha t) = R m_0 v$$

Therefore,

$$r(t) = \frac{R m_0 v e^{\lambda t}}{m_0(v + \alpha t)}$$

Note, since the mass is exponentially decreasing, the rope must increase its length exponentially to keep the angular momentum constant.

PROBLEM 3.9

a. Calculate the center of the mass of a uniform cone.
b. How does the center of the mass change if the density depends on the radius as $\rho = kr$, where k is a constant with appropriate units?

SOLUTION 3.9

a. The cone is positioned such that the apex is at the origin, and the axis of symmetry is in the z direction (Figure 3.6). Due to symmetry, the X and Y coordinates of the center of mass are zero.
   Only Z remains to be calculated

$$Z = \frac{1}{M} \int \rho z \, dV$$

From the similar triangles it follows that

$$\frac{r}{z} = \frac{R}{h}$$

So

$$r = \frac{R}{h} z$$

In cylindrical coordinates, $dV = r \, dr \, d\phi \, dz$ and $dV = \pi r^2 dz$ and r will be substituted by $\frac{R}{h} z$ from the previous equation, knowing that the density per mass is the inverse of the volume of the cone

$$\frac{\rho}{M} = \frac{3}{\pi R^2 h}$$

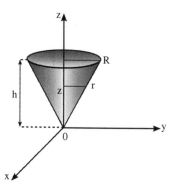

FIGURE 3.6 Inverted cone of height h and radius of the base R, with the apex centered at the origin of cartesian system.

$$Z = \frac{1}{M} \int \rho z \, \pi r^2 dz = \frac{1}{M} \int \rho \, \pi \, z \frac{R^2}{h^2} z^2 dz$$

$$= \frac{\pi \rho R^2}{h^2 M} \int_0^h z^3 dz = \frac{\pi R^2}{h^2} \frac{3}{\pi R^2 h} \frac{z^4}{4} \Big|_0^h = \frac{3}{4} h$$

Therefore, for a uniformly distributed mass,

$$X = 0, \quad Y = 0, \quad Z = \frac{3}{4} h$$

b. In this case, the density is not homogeneously distributed, but it depends on the radius as $\rho = kr$.

$$Z = \frac{1}{M} \int \rho z \, dV = \frac{1}{M} \int krz \, dV = \frac{1}{M} \int_0^h k \frac{R}{h} z z \pi \frac{R^2}{h^2} z^2 dz$$

$$= \frac{\pi k R^3}{Mh^3} \int_0^h z^4 dz = \frac{\pi k R^3}{Mh^3} \frac{z^5}{5} \Big|_0^h = \frac{\pi k R^3 h^2}{5M}$$

The mass of the cone is also dependent on the density:

$$M = \int \rho \, dV = \int kr \, \pi r^2 dz = \pi k \int_0^h \frac{z^3 R^3}{h^3} dz = \frac{\pi k R^3}{h^3} \frac{z^4}{4} \Big|_0^h = \frac{\pi k R^3 h}{4}$$

Back to the center of the mass,

$$Z = \frac{\pi k R^3 h^2}{5M} = \frac{\pi k R^3 h^2}{5} \frac{4}{\pi k R^3 h} = \frac{4}{5} h$$

Note that with $\rho = kr$ the center of the mass moves closer to the basis of the cone and further from the apex.

Also, note that the position of the center of mass depends on how we place the cone in the system of coordinates. In another problem, the cone is placed with the apex up, and the position of the center of mass is inverted, accordingly.

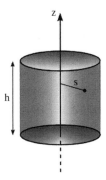

FIGURE 3.7 Uniform cylinder of height h.

PROBLEM 3.10

a. Calculate the moment of inertia of a uniform solid cylinder of radius R, rotating about its axis (Figure 3.7).
b. What is the moment of inertia if the density is not uniform, but proportional to the square of the radius $\rho = kr^2$, where k is a positive constant with appropriate units?
c. What is the moment of inertia if the density is $\rho = kr^n$, with k a constant with appropriate units and n natural number? Check your answer for $n = 2$.

SOLUTION 3.10

a. The definition of the moment of inertia is

$$I = \sum_{\alpha=1}^{N} m_\alpha s_\alpha^2$$

with s_α the distances of the mass m_α from the axis of rotation. Transform the sum into an integral of the form

$$I = \int \rho s^2 dV$$

The element of volume in cylindrical coordinates is $dV = r\, dr\, d\phi\, dz$.

$$I = \int \rho s^2 dV$$

$$I = \rho \int_0^R r^3 dr \int_0^{2\pi} d\phi \int_0^h dz = 2\pi\rho h \left.\frac{r^4}{4}\right|_0^R = \pi\rho h \frac{R^4}{2}$$

The density is represented in terms of mass and the volume as follows:

$$\rho = \frac{M}{V} = \frac{M}{\pi R^2 h}$$

The moment of inertia becomes

$$I = \pi \rho h \frac{R^4}{2} = \frac{\pi M h}{\pi R^2 h} \frac{R^4}{2} = \frac{MR^2}{2}$$

Note that this is the moment of inertia of a thin disk of radius R rotating about its axis.

b. For the case with the density $\rho = kr^2$, the density will remain under the integral sign, because it is not a constant anymore.

$$I = \int_0^R kr^2 r^3 dr \int_0^{2\pi} d\phi \int_0^h dz = 2\pi kh \left.\frac{r^6}{6}\right|_0^R = \pi kh \frac{R^6}{3}$$

c. The last case with the density $\rho = kr^n$ can be solved in the same manner

$$I = \int_0^R kr^n r^3 dr \int_0^{2\pi} d\phi \int_0^h dz = 2\pi kh \left.\frac{r^{n+4}}{n+4}\right|_0^R = 2\pi kh \frac{R^{n+4}}{n+4}$$

For $n = 2$, the previous expression from point (b) is obtained.

PROBLEM 3.11
Consider the cone in Figure 3.8 with its vertex at the origin and its axis on the z-axis. Given height h, radius R, and density $\rho(z) = \rho_0 \sin(\lambda z)$, find its center of mass.

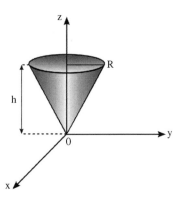

FIGURE 3.8 Solid cone of height h and base radius R.

SOLUTION 3.11

Since this is symmetric about the z-axis, it is only necessary to determine where the z-axis center of mass lies. This is given by

$$Z = \frac{1}{M} \int_v \rho z \, dV$$

where

$$dV = r \, d\phi \, dr \, dz$$

with

$$0 \le r \le \frac{R}{h} z, \, 0 \le \phi \le 2\pi, \, 0 \le z \le h$$

Therefore, the Z center of mass is

$$Z = \frac{1}{M} \int_0^h \int_0^{\frac{R}{h}z} \int_0^{2\pi} \rho_0 \sin(\lambda z) rz \, d\phi dr dz$$

$$= \frac{2\pi\rho_0}{M} \int_0^h \int_0^{\frac{R}{h}z} \sin(\lambda z) rz \, drdz = \frac{\pi\rho_0 R^2}{Mh^2} \int_0^h \sin(\lambda z) z^3 \, dz$$

$$= \frac{\pi\rho_0 R^2}{Mh^2} \left(\frac{h\lambda(6 - h^2\lambda^2)\cos(h\lambda) + 3(h^2\lambda^2 - 2)\sin(h\lambda)}{\lambda^4} \right)$$

and the complete center of mass is given by

$$\vec{R} = \left(0, 0, \frac{\pi\rho_0 R^2}{Mh^2\lambda^4} (h\lambda(6 - h^2\lambda^2)\cos(h\lambda) + 3(h^2\lambda^2 - 2)\sin(h\lambda)) \right)$$

PROBLEM 3.12

Consider a sphere centered at the origin as in Figure 3.9. Given a radius R and a density $\rho(\phi) = \rho_0 \sin(\lambda\phi)$, find the center of mass.

SOLUTION 3.12

The center of mass is given by

$$\vec{R} = \frac{1}{M} \int \rho \vec{r} \, dV$$

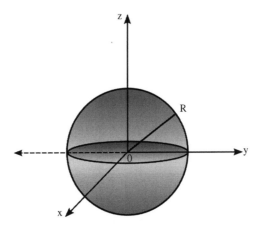

FIGURE 3.9   Solid sphere of radius R.

where $\vec{r} = x\,\hat{x} + y\,\hat{y} + z\,\hat{z}$ and $dV = r^2 \sin\theta \, d\phi \, d\theta \, dr$. Splitting this into components yields

$$X = \frac{1}{M} \int \rho_0 \sin(\lambda\phi)x \, dV = \frac{\rho_0}{M} \int_0^R \int_0^\pi \int_0^{2\pi} \sin(\lambda\phi)xr^2 \sin(\theta) \, d\phi \, d\theta \, dr$$

Expressing x in polar coordinates as $r\sin(\theta)\cos(\phi)$ yields

$$X = \frac{\rho_0}{M} \int_0^R \int_0^\pi \int_0^{2\pi} \sin(\lambda\phi)r^3 \sin^2(\theta)\cos(\phi) \, d\phi \, d\theta \, dr$$

Similarly, the other coordinates are given by

$$Y = \frac{\rho_0}{M} \int_0^R \int_0^\pi \int_0^{2\pi} \sin(\lambda\phi)yr^2 \sin(\theta) \, d\phi \, d\theta \, dr = \frac{\rho_0}{M} \int_0^R \int_0^\pi \int_0^{2\pi} \sin(\lambda\phi)r^3 \sin^2(\theta)\sin(\phi) \, d\phi \, d\theta \, dr$$

$$Z = \frac{\rho_0}{M} \int_0^R \int_0^\pi \int_0^{2\pi} \sin(\lambda\phi)zr^2 \sin(\theta) \, d\phi \, d\theta \, dr = \frac{\rho_0}{M} \int_0^R \int_0^\pi \int_0^{2\pi} \sin(\lambda\phi)r^3 \sin(\theta)\cos(\theta) \, d\phi \, d\theta \, dr$$

which use $y = r\sin(\theta)\sin(\phi)$ and $z = r\cos(\theta)$. The Z coordinate can be simplified by considering the ϕ integral only involves $\sin(\lambda\phi)$, which is zero. Therefore,

$$Z = 0$$

The X and Y integrals can be rearranged in the following way:

$$X = \frac{\rho_0}{M} \int_0^R r^3 \int_0^\pi \sin^2(\theta) \int_0^{2\pi} \sin(\lambda\phi)\cos(\phi) \, d\phi d\theta dr$$

$$Y = \frac{\rho_0}{M} \int_0^R r^3 \int_0^\pi \sin^2(\theta) \int_0^{2\pi} \sin(\lambda\phi)\sin(\phi) \, d\phi d\theta dr$$

which illustrate they are not terribly complicated. Therefore, the center of mass is given by

$$\vec{R} = \left(\frac{\pi R^4 \lambda \rho_0 \sin^2(\pi\lambda)}{4M(\lambda^2 - 1)}, \frac{\pi R^4 \rho_0 \sin^2(2\pi\lambda)}{8M(\lambda^2 - 1)}, 0 \right)$$

PROBLEM 3.13

Consider a mass m moving with velocity v colliding with another mass m attached to a spring with a spring constant k. Determine how much the spring compresses for

a. An inelastic collision (Figure 3.10).
b. An elastic collision at an angle θ (Figure 3.11) Assume the mass connected to the spring cannot deflect out of the $\pm \hat{x}$ direction.

FIGURE 3.10 Two masses colliding inelastically.

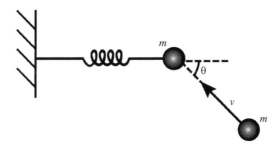

FIGURE 3.11 Two masses colliding elastically.

SOLUTION 3.13

a. The initial momentum is given by

$$\vec{p}_i = -mv\hat{x}$$

Since this is an inelastic collision and the masses are identical, the momentum at the moment when the mass hits the spring is

$$\vec{p}_f = -2mv_f\hat{x}$$

where v_f is determined via conservation of momentum

$$\vec{p}_i = \vec{p}_f$$

so

$$v_f = \frac{v}{2}$$

Therefore, the instant the spring starts compressing, the total mass is $2m$ and the velocity is $\frac{v}{2}$. The compression can be determined by considering the initial kinetic energy and final potential energy.

$$KE_i = U_f$$

$$\frac{1}{2}2mv_f^2 = \frac{1}{2}kx^2$$

$$x^2 = \frac{2m}{k}\frac{v^2}{4} = \frac{1}{2}\frac{m}{k}v^2$$

The compression is thus given by

$$x = \sqrt{\frac{m}{2k}}v$$

b. For the elastic collision, the diagram is as in Figure 3.12:
The initial momentum is given by

$$\vec{p}_i = mv(-\cos\theta\,\hat{x} + \sin\theta\,\hat{y})$$

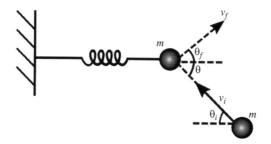

FIGURE 3.12 Geometry of the elastic collision.

Assuming the mass on the spring cannot deflect out of the $\pm \hat{x}$ direction, the final momentum is given by

$$\vec{p}_f = -mv_{f,s}\,\hat{x} + mv_f(\cos\theta_f\,\hat{x} + \sin\theta_f\,\hat{y})$$

Since this is an elastic collision, the kinetic energy before and after must be the same

$$KE_i = KE_f$$

$$\frac{1}{2}mv^2 = \frac{1}{2}mv_{f,s}^2 + \frac{1}{2}mv_f^2$$

The quantity of interest is $v_{f,s}$ as this will give the initial energy that goes into compressing the spring. Considering conservation of momentum, the following system of equations exists:

i. $-v\cos\theta = -v_{f,s} + v_f\cos\theta_f$, and yields that $v_f\cos\theta_f = v_{f,s} - v\cos\theta$
ii. $v\sin\theta = v_f\sin\theta_f$
iii. $v^2 = v_{f,s}^2 + v_f^2$

Taking (i) and (ii)

$$\frac{(ii)}{(i)}$$

$$\frac{v_f\sin\theta_f = v\sin\theta}{v_f\cos\theta_f = v_{f,s} - v\cos\theta}$$

$$\tan\theta_f = \frac{v\sin\theta}{v_{f,s} - v\cos\theta}$$

Equation (ii) now becomes

$$v_f = \frac{v\sin\theta}{\sin\left(\tan^{-1}\left(\dfrac{v\sin\theta}{v_{f,s} - v\cos\theta}\right)\right)}$$

Therefore, using (iii),

$$v_{f,s}^2 = v^2 - v_f^2$$

$$v_{f,s} = v^2 - \frac{v^2\sin^2\theta}{\sin^2\left(\tan^{-1}\left(\dfrac{v\sin\theta}{v_{f,s} - v\cos\theta}\right)\right)}$$

$$v_{f,s} = v\sqrt{1 - \frac{\sin^2\theta}{\sin^2\left(\tan^{-1}\left(\dfrac{v\sin\theta}{v_{f,s} - v\cos\theta}\right)\right)}}$$

Despite not being able to get a closed-form solution for $v_{f,s}$, this can be solved numerically for a given v and θ. As in part (a), the compression is given by

$$x^2 = \frac{m}{k}v_{f,s}^2$$

$$x = \sqrt{\frac{m}{k}}v_{f,s}$$

which can be determined using the numerical solution for $v_{f,s}$.

4 Energy

4.1 THEORY

This chapters introduces conservative forces, conditions for a force to be conservative, and law of conservation of energy.

4.1.1 WORK KINETIC ENERGY THEOREM

The variation of kinetic energy of a particle is equal to the work done by the net force acting on the particle

$$\Delta T = T_2 - T_1 = W$$

4.1.2 CONSERVATIVE FORCES

A force acting on a particle is conservative if it satisfies the following conditions:

i. The force depends only on the particle's position $\vec{r}$ and not on the velocity $\vec{v}$, the time t, or any other variables.
ii. The work done by the force is path independent.

Another condition for a force to be conservative is that the curl of the force should be zero. For example, if the force is represented in cartesian coordinate, the curl is

$$\nabla \times \vec{F} = \begin{vmatrix} \hat{x} & \hat{y} & \hat{z} \\ \dfrac{\partial}{\partial x} & \dfrac{\partial}{\partial y} & \dfrac{\partial}{\partial z} \\ F_x & F_y & F_z \end{vmatrix} = 0$$

It is important to note that the potential energy can be defined only for conservative fields.

The relationship between the force and the potential energy is that the force is minus the gradient of the potential energy:

$$\vec{F} = -\nabla U$$

If the potential energy is known, the force can be obtained.

DOI: 10.1201/9781003365709-4

If the force is known, it is important to verify that the force is conservative (the curl is zero), and then the potential energy is calculated as

$$U = -\int \vec{F} \cdot d\vec{r}$$

Principle of conservation of mechanical energy: if the forces acting on a particle are conservative, the mechanical energy of the particle is conserved.

The total mechanical energy is conserved only when the potential energy is time independent, that is, $\dfrac{\partial U}{\partial t} = 0$.

4.1.3   OBTAINING THE EQUATION OF THE MOTION FROM THE CONSERVATION OF THE ENERGY

In one-dimensional example, the mechanical energy is the sum of the kinetic energy and potential energy,

$$E = T + U(x)$$

The kinetic energy has the form

$$T = \frac{m\dot{x}^2}{2}$$

The kinetic energy is substituted into the energy equation and expressions for velocity and position can be obtained:

$$E = \frac{m\dot{x}^2}{2} + U(x)$$

$$\dot{x}(x) = \pm\sqrt{\frac{2}{m}(E - U(x))}$$

From here, $\dot{x} = \dfrac{dx}{dt}$ and interestingly, the time is obtained as a function of position, and subsequently, the position versus time can be calculated

$$dt = \frac{dx}{\dot{x}(x)}$$

By integration,

$$t - t_0 = \int_{x_0}^{x} \frac{dx}{\dot{x}(x)} = \sqrt{\frac{m}{2}} \int_{x_0}^{x} \frac{dx}{\sqrt{E - U(x)}}$$

From here time as a function of position is obtained, and then position as a function of time can be calculated.

4.2 PROBLEMS AND SOLUTIONS

PROBLEM 4.1

Check if the following force is conservative $\vec{F} = (ax, bx, cz)$. If it is conservative, find the potential energy and verify that $\vec{F} = -\nabla U$.

SOLUTION 4.1

One condition for a force to be conservative is to have the curl equal to zero

$$\nabla \times \vec{F} = \begin{vmatrix} \hat{x} & \hat{y} & \hat{z} \\ \dfrac{\partial}{\partial x} & \dfrac{\partial}{\partial y} & \dfrac{\partial}{\partial z} \\ F_x & F_y & F_z \end{vmatrix} = \begin{vmatrix} \hat{x} & \hat{y} & \hat{z} \\ \dfrac{\partial}{\partial x} & \dfrac{\partial}{\partial y} & \dfrac{\partial}{\partial z} \\ ax & by & cz \end{vmatrix}$$

$$\nabla \times \vec{F} = \hat{x}\left(\frac{\partial(cz)}{\partial y} - \frac{\partial(by)}{\partial z}\right) + \hat{y}\left(\frac{\partial(ax)}{\partial z} - \frac{\partial(cz)}{\partial x}\right) + \hat{z}\left(\frac{\partial(by)}{\partial x} - \frac{\partial(ax)}{\partial y}\right)$$

$$\nabla \times \vec{F} = (0, 0, 0)$$

$$\nabla \times \vec{F} = \vec{0}$$

This force is conservative and the potential energy can be calculated

$$U = -\int \vec{F} \cdot d\vec{r} = -\int F_x \, dx + F_y \, dy + F_z \, dz = -\left(\int ax \, dx + \int by \, dy + \int cz \, dz\right)$$

$$= -\left(\frac{ax^2}{2} + \frac{by^2}{2} + \frac{cz^2}{2}\right) + \text{Constant}$$

Since the potential energy is given up to a constant, the constant may be chosen as zero.

Now, it can be verified that the force is the minus gradient of the potential energy, $\vec{F} = -\nabla U$ by a simple derivation after writing the gradient in cartesian coordinates

$$\vec{F} = -\nabla U = -\left(\frac{\partial U}{\partial x}\hat{x} + \frac{\partial U}{\partial y}\hat{y} + \frac{\partial U}{\partial z}\hat{z}\right) = ax\,\hat{x} + by\,\hat{y} + cz\,\hat{z} = (ax, bx, cz) = \vec{F}$$

PROBLEM 4.2

Check if the following forces are conservative. If they are, find the potential energy.

a. $\vec{F} = (x^2 + x, 2y - 1, 0)$

b. $\vec{F} = (x^2 + x, 2y - z, x^3 + y/2)$

c. $\vec{F} = (2x + y, x + 2z, 2y + z)$

SOLUTION 4.2

If the curl of a force is 0, then the force is conservative.

a.

$$\nabla \times \vec{F} = \begin{vmatrix} \hat{x} & \hat{y} & \hat{z} \\ \partial/\partial x & \partial/\partial y & \partial/\partial z \\ x^2 + x & 2y - 1 & 0 \end{vmatrix}$$

$$\nabla \times \vec{F} = \left(\frac{\partial}{\partial y}(0) - \frac{\partial}{\partial z}(2y - 1), \frac{\partial}{\partial z}(x^2 + x) - \frac{\partial}{\partial x}(0), \frac{\partial}{\partial x}(2y - 1) - \frac{\partial}{\partial y}(x^2 + x) \right)$$

$$\nabla \times \vec{F} = (0, 0, 0)$$

$$\nabla \times \vec{F} = \vec{0}$$

This force is conservative. The potential energy can be found

$$U = -\int \vec{F} \cdot d\vec{r}$$

$$U = -\left(\frac{x^3}{3} + \frac{x^2}{2} + y^2 - y \right)$$

b.

$$\nabla \times \vec{F} = \begin{vmatrix} \hat{x} & \hat{y} & \hat{z} \\ \partial/\partial x & \partial/\partial y & \partial/\partial z \\ x^2 + x & 2y - z & x^3 + \frac{y}{2} \end{vmatrix}$$

$$\nabla \times \vec{F} = \left(\frac{\partial}{\partial y}\left(x^3 + \frac{y}{2} \right) - \frac{\partial}{\partial z}(2y - z), \frac{\partial}{\partial z}(x^2 + x) - \frac{\partial}{\partial x}\left(x^3 + \frac{y}{2} \right), \right.$$

$$\left. \frac{\partial}{\partial x}(2y - z) - \frac{\partial}{\partial y}(x^2 + x) \right)$$

$$\nabla \times \vec{F} = (3/2, -3x^2, 0)$$

This force is not conservative.

c.

$$\nabla \times \vec{F} = \begin{vmatrix} \hat{x} & \hat{y} & \hat{z} \\ \partial/\partial x & \partial/\partial y & \partial/\partial z \\ 2x+y & x+2z & 2y+z \end{vmatrix}$$

$$\nabla \times \vec{F} = \left(\frac{\partial}{\partial y}(2y+z) - \frac{\partial}{\partial z}(x+2z), \frac{\partial}{\partial z}(2x+y) - \frac{\partial}{\partial x}(2y+z), \right.$$
$$\left. \frac{\partial}{\partial x}(x+2z) - \frac{\partial}{\partial y}(2x+y) \right)$$

$$\nabla \times \vec{F} = (0,0,0)$$

$$\nabla \times \vec{F} = \vec{0}$$

This force is conservative. The potential energy can be found

$$U = -\int \vec{F} \cdot d\vec{r}$$

$$U = -\left(x^2 + xy + 2yz + \frac{z^2}{2} \right)$$

PROBLEM 4.3
Evaluate the line integral for the work done by the following forces on both paths shown. Determine if each force is conservative.

Path a is along the line $y = x$ (Figure 4.1a).

Path b is along the line $y = x^2$ (Figure 4.1b).

a. $F_1 = (y^2, x)$
b. $F_2 = (2x, y)$

SOLUTION 4.3
a. Path a: The work done is $\int_a \vec{F}_1 \cdot d\vec{r}$.

$$W_a = \int_a F_{1x} dx + F_{1y} dy$$

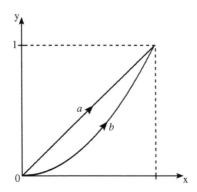

FIGURE 4.1 Path a (first bisector) and b (parabola).

On path a, $x = y$. It follows that $dx = dy$.

$$W_a = \int_a (F_{1x} + F_{1y})dx$$

$$W_a = \int_0^1 (y^2 + x)dx$$

$$W_a = \int_0^1 (x^2 + x)dx$$

$$W_a = \frac{5}{6}$$

Path b: The work done is $\int_b \vec{F}_1 \cdot d\vec{r}$

$$W_b = \int_b F_{1x}dx + F_{1y}dy$$

On path b, $y = x^2$. It follows that $dy = 2xdx$

$$W_b = \int_b F_{1x}dx + F_{1y}2xdx$$

$$W_b = \int_0^1 (y^2 + x \cdot 2x)dx$$

$$W_b = \int_0^1 (x^4 + 2x^2)dx$$

$$W_b = \frac{13}{15}$$

Since $W_a \neq W_b$, the force is not conservative.

b. Path a: The work done is

$$W_a = \int_a \vec{F_2} \cdot d\vec{r} = \int_a F_{2x}dx + F_{2y}dy$$

On path a, $x = y$. It follows that $dx = dy$

$$W_a = \int_a (F_{2x} + F_{2y})dx$$

$$W_a = \int_0^1 (2x + y)dx$$

$$W_a = \int_0^1 (2x + x)dx$$

$$W_a = \frac{3}{2}$$

Path b: The work done is

$$W_b = \int_b \vec{F_2} \cdot d\vec{r} = \int_b F_{2x}dx + F_{2y}dy$$

On path b, $y = x^2$. It follows that $dy = 2xdx$

$$W_b = \int_b F_{2x}dx + F_{2y}2xdx$$

$$W_b = \int_0^1 (2x + y \cdot 2x)dx$$

$$W_b = \int_0^1 (2x + 2x^3)dx$$

$$W_b = \frac{3}{2}$$

Since $W_a = W_b$, the force is conservative.

PROBLEM 4.4

Find the force corresponding to the following potential energy $U = -\dfrac{k}{r}$, with k a constant with appropriate units, then check that the force is conservative (it should be).

SOLUTION 4.4

The form of the potential energy suggests that the spherical coordinates may be indicated in this case. The potential energy depends only on the radius, not on azimuthal angle or polar angle.

$$\vec{F} = -\nabla U = -\left(\hat{r}\frac{\partial U}{\partial r} + \hat{\theta}\frac{1}{r}\frac{\partial U}{\partial \theta} + \hat{\phi}\frac{1}{r\sin\theta}\frac{\partial U}{\partial \phi} \right) = -\hat{r}\frac{\partial\left(-\dfrac{k}{r}\right)}{\partial r} = -\hat{r}\frac{k}{r^2}$$

The minus sign in the potential indicates that the force is attractive.

The force has a potential energy associated with it, so it must be a conservative force, therefore, the curl of the force should be zero. The curl in spherical coordinates is written as

$$\nabla \times \vec{F} = \hat{r}\frac{1}{r\sin\theta}\left(\frac{\partial}{\partial\theta}(\sin\theta\, F_\phi) - \frac{\partial}{\partial\phi}F_\theta \right) + \hat{\theta}\left(\frac{1}{r\sin\theta}\frac{\partial}{\partial\phi}F_r - \frac{1}{r}\frac{\partial}{\partial r}(rF_\phi) \right)$$

$$+ \hat{\phi}\frac{1}{r}\left(\frac{\partial}{\partial r}(rF_\theta) - \frac{\partial}{\partial\theta}F_r \right)$$

$$= \hat{r}\frac{1}{r\sin\theta}\left(\frac{\partial}{\partial\theta}(0) - \frac{\partial}{\partial\phi}0 \right) + \hat{\theta}\left(\frac{1}{r\sin\theta}\frac{\partial}{\partial\phi}\left(-\frac{k}{r^2}\right) - \frac{1}{r}\frac{\partial}{\partial r}(0) \right)$$

$$+ \hat{\phi}\frac{1}{r}\left(\frac{\partial}{\partial r}(0) - \frac{\partial}{\partial\theta}\left(-\frac{k}{r^2}\right) \right) = 0$$

Therefore, the force is conservative.

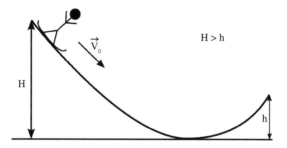

FIGURE 4.2 Skier going down ski jump ramp.

PROBLEM 4.5

A skier, starting with a velocity v_0, slides down a frictionless ski jump ramp as pictured in Figure 4.2. Using conservation of energy, find

a. The maximum speed of the skier.
b. The speed of the skier at the end of the ramp.

SOLUTION 4.5

a. The mechanical energy is $E = T + U$. At the top of the ramp, $E_1 = \frac{1}{2}mv_0^2 + mgH$. At the lowest point of the ramp, $E_2 = \frac{1}{2}mv_{max}^2$. Because of the conservation of energy, $E_1 = E_2$. Thus,

$$\frac{1}{2}mv_{max}^2 = \frac{1}{2}mv_0^2 + mgH$$

Solving for v_{max}, the maximum speed of the skier is

$$v_{max} = \sqrt{v_0^2 + 2gH}$$

b. At the end of the ramp, $E_3 = \frac{1}{2}mv_{end}^2 + mgh$. Because of conservation of energy, $E_1 = E_3$.

$$\frac{1}{2}mv_{end}^2 + mgh = \frac{1}{2}mv_0^2 + mgH$$

Solving for v_{end}, the speed of the skier at the end of the ramp is

$$v_{end} = \sqrt{v_0^2 + 2g(H - h)}$$

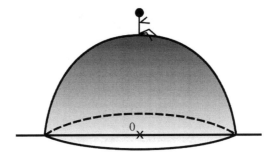

FIGURE 4.3   Child sitting on a perfectly hemispherical igloo.

PROBLEM 4.6

A child sits on an igloo. Find the vertical height at which the child leaves the surface of the igloo. Consider the igloo a perfect, frictionless hemisphere of radius R. The child's initial position is on the top of the igloo and consider that the slide is starting with a tiny nudge (Figure 4.3).

SOLUTION 4.6

This is a classical problem which can be solved by using the conservation of total mechanical energy and by considering that the motion occurs on a circular trajectory, with the total force being equal to the centripetal force. Also, at the moment the child leaves the igloo, the normal to the surface is zero.

The potential energy level is zero at the ground level. The potential energy depends on the angle θ as $U(\theta) = mgR\cos\theta$.

The total energy is conserved:

$$E = U(0) = mgR = \frac{mv^2}{2} + mgR\cos\theta$$

From here, the kinetic energy is obtained:

$$T = \frac{mv^2}{2} = E - U = mgR - mgR\cos\theta = mgR(1 - \cos\theta)$$

During the motion on the sphere, Newton's Second Law is written (Figure 4.4)

$$N - mg\cos\theta = -\frac{mv^2}{R}$$

From the energy conservation equation,

$$mv^2 = 2mgR(1 - \cos\theta)$$

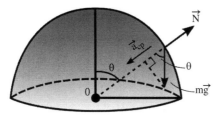

FIGURE 4.4 Child sliding on the hemispherical igloo, with the gravitational force and normal force represented. The centripetal acceleration is keeping the child on the igloo, until the point when the normal force becomes zero, moment in which the child falls from the igloo.

And back in Newton's Second Law,

$$N = mg\cos\theta - 2mg(1-\cos\theta) = mg(3\cos\theta - 2)$$

At the moment the child leaves the igloo, the normal force becomes zero, so the angle θ can be obtained as

$$N = mg(3\cos\theta - 2) = 0$$

So

$$\cos\theta = \frac{2}{3}$$

And therefore,

$$y = R\cos\theta = \frac{2}{3}R$$

PROBLEM 4.7

A uniform sphere of mass M and radius R is rolling down an incline of height h (Figure 4.5). Find the speed of the sphere at the base of the incline. Solve this problem in two ways.

a. First, consider the motion of the sphere about the center of the mass, with the kinetic energy of the center of the mass and the rotational energy about the center of the mass. Use the non-slip condition: $\omega = \dfrac{v}{R}$.

b. Consider the rotation about the point where the sphere touches the incline. For this, consider Steiner's theorem $I' = I + MR^2$, where is the moment of inertia of the sphere with respect to an axis going through the center of the mass $I = \dfrac{2MR^2}{5}$. Calculate the speed of the sphere at the base of the incline. Use the non-slip condition: $\omega = \dfrac{v}{R}$.

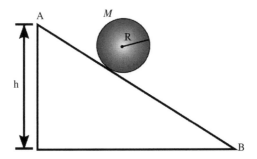

FIGURE 4.5   Sphere sliding on the incline.

SOLUTION 4.7

a. Here, the total mechanical energy is conserved.

$$E_A = E_B$$

On the top of the incline, the sphere starts with zero kinetic energy, but the potential energy is maximum:

$$E_A = Mgh$$

At the bottom of the incline, the total energy is only kinetic – we consider the lowest potential energy at the bottom of the incline

$$E_B = \frac{Mv^2}{2} + \frac{I\omega^2}{2}$$

where I is the moment of inertia of the sphere with respect to the central axis, $I = \dfrac{2MR^2}{5}$, and by using the non-slip condition and the conservation of the energy,

$$Mgh = \frac{Mv^2}{2} + \frac{I\omega^2}{2}$$

$$Mgh = \frac{Mv^2}{2} + \frac{\dfrac{2MR^2}{5}\left(\dfrac{v}{R}\right)^2}{2} = \frac{Mv^2}{2} + \frac{Mv^2}{5}$$

By rearranging the terms and dividing by the mass, the speed of the center of the mass of the sphere is obtained:

$$v = \sqrt{\frac{10gh}{7}}$$

b. Considering the energy with respect to the contact point, and by using the conservation of the energy,

$$\Delta T = -\Delta U$$

$$\Delta T = \frac{I'\omega^2}{2} = \frac{(I + MR^2)\left(\dfrac{v}{R}\right)^2}{2} = \frac{\left(\dfrac{2MR^2}{5} + MR^2\right)\left(\dfrac{v}{R}\right)^2}{2} = \frac{7}{10}Mv^2$$

Note that the moment of inertia I' is calculated using Steiner's theorem. As before, the potential energy is

$$\Delta U = 0 - Mgh$$

And from the conservation of energy,

$$\frac{7}{10}Mv^2 = Mgh$$

The speed of the sphere at the bottom of the incline is, as before,

$$v = \sqrt{\frac{10gh}{7}}$$

PROBLEM 4.8

A block is attached to a horizontal spring of constant k. There is a coefficient of friction μ between the block and the surface. The block is pulled and released at position x_s, as in Figure 4.6. Find the velocity of the block when it reaches the equilibrium position the first time, using energy conservation.

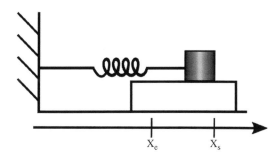

FIGURE 4.6 Mass connected to a spring released from rest at x_s.

SOLUTION 4.8

The work due to friction is $W_f = -fd = -\mu mgd$. The change in mechanical energy is due to friction, so $\Delta E = \Delta T + \Delta U = W_f$. At the equilibrium point,

$$-\frac{1}{2}mv_s^2 + \frac{1}{2}mv_e^2 - \frac{1}{2}kx_s^2 + \frac{1}{2}kx_e^2 = -\mu mg(x_s - x_e)$$

The origin of the axis is chosen to be the equilibrium point, and recall that $v_s = 0$.

$$\frac{1}{2}mv_e^2 - \frac{1}{2}kx_s^2 = -\mu mgx_s$$

Solving for v_e, the speed at equilibrium is

$$v_e = \sqrt{\frac{k}{m}x_s^2 - 2\mu gx_s}$$

PROBLEM 4.9

A ball of mass m is sliding up a frictionless ramp of length L and at an angle θ. Determine the maximum initial velocity of the ball such that it doesn't slide off the ramp. Consider the following two situations and compare the answers:

a. The ball is initially released parallel to the ramp, as in Figure 4.7.
b. The ball is initially released parallel to the floor, as in Figure 4.8.

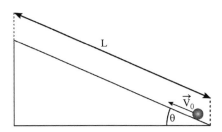

FIGURE 4.7　Ball traveling up frictionless ramp.

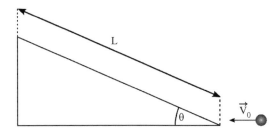

FIGURE 4.8　Ball traveling toward frictionless ramp.

Note: This is the same problem text as Problem 1.6, but now should be solved using energy considerations.

SOLUTION 4.9

a. Conservation of energy requires

$$E_i = E_f$$

where E_i is pure kinetic energy up the ramp and E_f is pure potential at the top of the ramp. Therefore,

$$\frac{1}{2} m v_0^2 = mgh$$

By dividing it by the mass and writing the height as $h = L\sin\theta$,

$$\frac{1}{2} v_0^2 = gL\sin\theta$$

This can be immediately solved

$$v_0 = \sqrt{2gL\sin\theta}$$

b. This will have a similar approach as part a, but care must be taken when the puck first interacts with the ramp. Initially,

$$E_1 = \frac{1}{2} m v_0^2$$

but after the ramp, the energy becomes

$$E_2 = \frac{1}{2} m v_0^2 \cos^2\theta$$

It is this E_2 that will be used in the conservation of energy equation, similar to part a, $E_i = E_f$.

$$\frac{1}{2} m v_0^2 \cos^2\theta = mgh$$

As before, after simplification of mass and substitution of height h,

$$\frac{1}{2} v_0^2 \cos^2\theta = gL\sin\theta$$

Therefore, the velocity is given by

$$v_0 = \frac{1}{\cos\theta}\sqrt{2gL\sin\theta}$$

as was found in Chapter 1.

PROBLEM 4.10

Consider a block of mass m sliding down a frictionless ramp at an incline θ as in Figure 4.9. Find the velocity of the block at time t if the block is stationary at $t = 0$.

Note: This is the same problem text as Problem 1.4, but now should be solved using energy considerations.

SOLUTION 4.10

Considering conservation of energy, the total energy of each of the following positions must be the same.

Specifically,

$$E_0 = E(t)$$

Since the block is initially at rest, at time t, the difference in potential energy has been converted to kinetic energy (Figure 4.10). Therefore,

$$mg\Delta h = \frac{1}{2}mv^2$$

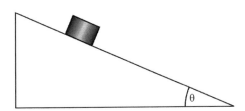

FIGURE 4.9   Block sliding down a ramp.

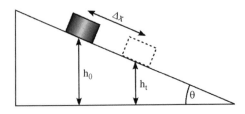

FIGURE 4.10   The displacement of the block as it slides down the ramp.

The change in height is given by $\Delta h = \Delta x \sin\theta$. Considering an infinitesimal displacement, Δx becomes $dx = v\,dt$. From this, the equation from conservation of energy is written as

$$gv\sin\theta\,dt = \frac{1}{2}v^2$$

and the velocity is

$$v = 2g\sin\theta\,dt$$

The solution $v = 0$ is not interesting, being the point up the ramp the object starts moving from. The expression for velocity can be directly integrated, yielding

$$v(t) = 2gt\sin\theta + C$$

where C is a constant. Since the initial velocity at $t = 0$ is zero, $C = 0$ and the velocity of the block is

$$v(t) = 2gt\sin\theta$$

Precisely what was found in Chapter 1.

PROBLEM 4.11

In Problem 2.8, a mass on a pendulum moved in the presence of linear and quadratic air resistance and the equations of motion were found. For the linear drag case, and assuming small angles, find an expression for total energy as a function of time. Assume the initial position is ϕ_0 and initial velocity is zero.

SOLUTION 4.11

From Problem 2.8, the equation of motion was found to be

$$\ddot{\phi} + \frac{b}{m}\dot{\phi} + \frac{g}{l}\sin\phi = 0$$

In the small angle approximation, this becomes

$$\ddot{\phi} + \frac{b}{m}\dot{\phi} + \frac{g}{l}\phi = 0$$

This differential equation can be solved considering the characteristic equation

$$r^2 + \frac{b}{m}r + \frac{g}{l} = 0$$

which has roots

$$r_\pm = -\frac{b}{2m} \pm \sqrt{\frac{b^2}{4m^2} - \frac{g}{l}}$$

Therefore, the position is given by

$$\phi(t) = Ae^{r_+ t} + Be^{r_- t}$$

where A and B are constants. Also,

$$\dot{\phi}(t) = Ar_+ e^{r_+ t} + Br_- e^{r_- t}$$

Using the initial conditions,

$$\dot{\phi}(0) = Ar_+ + Br_- = 0 \text{ and } B = -A\frac{r_+}{r_-}$$

The position at $t = 0$ is

$$\phi(0) = A + B = \phi_0$$

And by substituting B,

$$A - A\frac{r_+}{r_-} = \phi_0 \text{ from where } \frac{A}{r_-}(r_- - r_+) = \phi_0$$

The constants are

$$A = \frac{\phi_0 r_-}{(r_- - r_+)}$$

so

$$B = -\frac{\phi_0 r_+}{(r_- - r_+)}$$

and the position is given by

$$\phi(t) = \frac{\phi_0}{(r_- - r_+)}(r_- e^{r_+ t} - r_+ e^{r_- t})$$

The total energy is the sum of the kinetic and potential energies so

$$E(t) = KE(t) + U(t) = \frac{1}{2}mv^2 + mgh$$

The velocity is given by

$$v = l\dot\phi = l\frac{\phi_0}{(r_- - r_+)}(r_- r_+ e^{r_+ t} - r_+ r_- e^{r_- t}) = l\frac{\phi_0 r_- r_+}{(r_- - r_+)}(e^{r_+ t} - e^{r_- t})$$

and height of the mass is given by

$$h = l(1 - \cos\phi)$$

In the small angle approximation,

$$h \approx l\left(1 - \left(1 - \frac{\phi^2}{2}\right)\right) = \frac{1}{2}l\phi^2$$

Therefore,

$$E(t) = \frac{1}{2}ml^2\dot\phi^2(t) + \frac{1}{2}mgl\,\phi^2(t)$$

$$= \frac{1}{2}ml\left(1\frac{\phi_0^2 r_-^2 r_+^2}{(r_- - r_+)^2}(e^{r_+ t} - e^{r_- t})^2 + g\frac{\phi_0^2}{(r_- - r_+)^2}(r_- e^{r_+ t} - r_+ e^{r_- t})^2\right)$$

$$= \frac{\phi_0^2 ml}{2(r_- - r_+)^2}(lr_-^2 r_+^2 (e^{r_+ t} - e^{r_- t})^2 + g(r_- e^{r_+ t} - r_+ e^{r_- t})^2)$$

To understand this motion, it is necessary to examine $e^{r_\pm}$

$$e^{r_\pm t} = e^{\left(-\frac{b}{2m} \pm \sqrt{\frac{b^2}{4m^2} - \frac{g}{l}}\right)t} = e^{-\frac{b}{2m}t}e^{\pm t\sqrt{\frac{b^2}{4m^2} - \frac{g}{l}}}$$

Notice

$$\frac{b}{2m} = \sqrt{\frac{b^2}{4m^2}} > \sqrt{\frac{b^2}{4m^2} - \frac{g}{l}}$$

Therefore, the $e^{-\frac{b}{2m}t}$ part of the exponential dominates and the motion of the pendulum exponentially decays. For more insight, consider the other part of the exponential

$$e^{\pm t\sqrt{\frac{b^2}{4m^2} - \frac{g}{l}}}$$

When $\frac{b^2}{4m^2} > \frac{g}{l}$, $\frac{b^2}{4m^2} - \frac{g}{l} > 0$, and $\sqrt{\frac{b^2}{4m^2} - \frac{g}{l}}$ is real. In this case, damping is large, and the mass essentially swing down to its lowest point and stops. This may be the situation if the pendulum was in a viscous liquid. When $\dfrac{b^2}{4m^2} < \dfrac{g}{l}$, it follows that $\dfrac{b^2}{4m^2} - \dfrac{g}{l} < 0$ and $\sqrt{\dfrac{b^2}{4m^2} - \dfrac{g}{l}}$ is complex (yielding a complex exponential). In this case, damping is small, and the mass oscillates around the lowest point with the amplitude of oscillation being exponential damped, in accordance with $e^{-\frac{b}{2m}t}$.

PROBLEM 4.12

A ball is placed at the top of a frictionless ramp with a height h and incline θ (Figure 4.11). The ball is also attached to a point $2h$ above the ground (h above the top of the ramp) with a string of length $2h$ and negligible mass. Find the initial velocity of the ball must have, parallel to the ramp, in order to reach a maximum height halfway between the fixed point and the top of the ramp.

Note: This is the same problem text as Problem 1.15 but now should be solved using energy considerations.

SOLUTION 4.12

Considering the two separate motions (down the ramp vs. leaving the ramp and ascending to $2h$), there are several energies involved. There is the initial kinetic energy associated with the initial velocity and the initial potential energy associated with being at the top of a ramp. There is the intermediate energy associated with the moment the ball leaves the ramp where the initial potential energy has been converted to kinetic energy. Then there is the final energy of the ball at its final height where all the kinetic energy has been converted to potential energy.

To help keep the goal of the problem in mind, consider working "backward" and looking at the energies in the second phase of the motion. Here, the final energy is purely potential and the initial is purely kinetic. Since the velocity here is the intermediate velocity mentioned above, it is denoted as v_{1y}. Therefore,

$$\frac{1}{2}mv_{1y}^2 = mgH$$

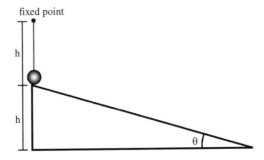

FIGURE 4.11   The ball positioned at the top of a ramp, connected to a fixed point.

Since the exact spot the ball leaves the ramp is unknown, it can be taken to be a distance y beneath the top of the ramp. This means $H = h + y$. Therefore,

$$\frac{1}{2}v_{1y}^2 = g(h + y)$$

The height h is known, the distance y is unknown but can be determined by geometry, and the velocity v_{1y} is unknown but can be determined by the first phase of the motion. Consider Figure 4.12 to determine y.

The equations then follow from the triangle depicted in Figure 4.13.

$$\tan\theta = \frac{y}{x} \text{ so } x = \frac{y\cos\theta}{\sin\theta}$$

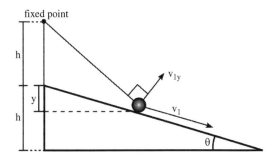

FIGURE 4.12 The ball moving as far down the ramp as possible before it must leave the surface of the ramp.

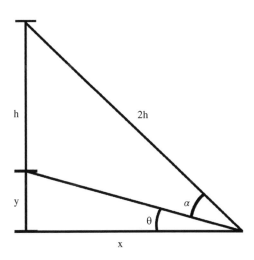

FIGURE 4.13 Two triangles depicting the angles and heights immediately before the ball leaves the ramp.

From the Pythagorean theorem in the big triangle,

$$x^2 + (h+y)^2 = (2h)^2$$

$$\frac{y^2 \cos^2 \theta}{\sin^2 \theta} + h^2 + 2hy + y^2 = 4h^2$$

Rearranging the terms, it follows that

$$\frac{y^2}{\sin^2 \theta}(\cos^2 \theta + \sin^2 \theta) + 2hy - 3h^2 = 0$$

And after recalling that $\cos^2 \theta + \sin^2 \theta = 1$,

$$y^2 + 2h\sin^2 \theta y - 3h^2 \sin^2 \theta = 0$$

From this,

$$y = \frac{-2h\sin^2 \theta \pm \sqrt{4h^2 \sin^4 \theta - 4\left(-3h^2 \sin^2 \theta\right)}}{2} = h\sin\theta\left(\sqrt{\sin^2 \theta + 3} - \sin\theta\right)$$

To determine the velocity v_{1y}, consider Figure 4.14.

$$v_{1y} = v_1 \cos\gamma$$

with $\alpha + \gamma = \dfrac{\pi}{2}$ so $\gamma = \dfrac{\pi}{2} - \alpha$

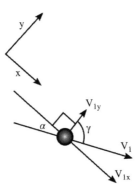

FIGURE 4.14 Detailed illustration of the forces with respect to the selected axes.

$$v_{1y} = v_1 \cos\left(\frac{\pi}{2} - \alpha\right) = v_1 \sin\alpha$$

It also follows that $\sin(\alpha + \theta) = \dfrac{h+y}{2h}$

$$\alpha = \sin^{-1}\left(\frac{h+y}{2h}\right) - \theta = \sin^{-1}\left(\frac{1 + \sin\theta\left(\sqrt{\sin^2\theta + 3} - \sin\theta\right)}{2}\right) - \theta$$

Thus,

$$v_{1y} = v_1 \sin\left(\sin^{-1}\left(\frac{\sin\theta\left(\sqrt{\sin^2\theta + 3} - \sin\theta\right)}{2}\right) - \theta\right)$$

Using the conservation of energy equation above, $\dfrac{1}{2}v_{1y}^2 = g(h+y)$

$$v_{1y} = \sqrt{2gh\left(1 + \sin\theta\left(\sqrt{\sin^2\theta + 3} - \sin\theta\right)\right)}$$

Therefore,

$$v_1 = \frac{\sqrt{2gh\left(1 + \sin\theta\left(\sqrt{\sin^2\theta + 3} - \sin\theta\right)\right)}}{\sin\left(\sin^{-1}\left(\dfrac{\sin\theta\left(\sqrt{\sin^2\theta + 3} - \sin\theta\right)}{2}\right) - \theta\right)}$$

Now is to obtain v_1 from the first phase of the motion. Recall, this is the initial kinetic and potential energies being converted to kinetic energy

$$\frac{1}{2}mv_0^2 + mgy = \frac{1}{2}mv_1^2$$

with v_0 being the initial velocity (and the objective of the problem). This can be directly solved for v_0

$$v_0 = \sqrt{v_1^2 - 2gy}$$

$$= \sqrt{\sin^2\left(\sin^{-1}\left(\frac{\sin\theta\left(\sqrt{\sin^2\theta + 3} - \sin\theta\right)}{2}\right) - \theta\right)} \left[\frac{2gh\left(1 + \sin\theta\left(\sqrt{\sin^2\theta + 3} - \sin\theta\right)\right)}{} - 2gh\sin\theta\left(\sqrt{\sin^2\theta + 3} - \sin\theta\right) \right]}$$

Slightly different geometry was used here versus that in Problem 1.15, but from the general structure, it is easy to see the similarities with the previous solution. The reorganization can be achieved via trigonometric identities.

5 Oscillations

5.1 THEORY

This chapters introduces Hooke's Law, the simple harmonic motion and different types of oscillations as damped, or driven damped oscillations. The parallelism between mechanical and electrical oscillators is discussed.

5.1.1 HOOKE'S LAW

The force exerted by a spring is

$$F(x) = -kx$$

where k is the spring constant (or force constant, in N/m) and x is the distance with respect to the equilibrium point.

The potential energy is

$$U(x) = \frac{kx^2}{2}$$

The potential energy can be expanded in a Taylor series as

$$U(x) = U(0) + U'(0)x + \frac{U''(0)\, x^2}{2} + \cdots$$

5.1.2 SIMPLE HARMONIC MOTION

Starting with the force, and expressing the force by using Newton's Second Law,

$$F(x) = -kx$$

$$F(x) = m\ddot{x}$$

From this, the equation of motion is

$$m\ddot{x} = -kx$$

$$\ddot{x} = -\frac{k}{m}x = -\omega^2 x$$

where ω is the angular frequency of oscillation.

DOI: 10.1201/9781003365709-5

This is a second-order, linear, homogeneous differential equation, with two independent solutions, which can be written in different ways, for example:

$$x(t) = e^{i\omega t} \text{ and } x(t) = e^{-i\omega t}$$

The linear combination of the two solutions is also a solution,

$$x(t) = C_1 e^{i\omega t} + C_2 e^{-i\omega t}$$

After writing

$$e^{\pm i\omega t} = \cos \omega t \pm i \sin \omega t$$

$$x(t) = (C_1 + C_2)\cos \omega t + i(C_1 - C_2)\sin \omega t = B_1 \cos \omega t + B_2 \sin \omega t$$

Which is the form of simple harmonic motion.

5.1.3 ENERGY

Potential energy is written as

$$U = \frac{kx^2}{2}$$

Kinetic energy as

$$T = \frac{m\dot{x}^2}{2}$$

If we consider a third form for the simple harmonic motion,

$$x(t) = A \cos(\omega t - \delta)$$

Then the total energy is

$$E = T + U = \frac{m\omega^2 A^2}{2} \sin^2(\omega t - \delta) + \frac{kA^2}{2} \cos^2(\omega t - \delta) = \frac{kA^2}{2}$$

5.1.4 PARTICULAR TYPES OF OSCILLATIONS AND THE DIFFERENTIAL EQUATIONS ASSOCIATED WITH THEM

5.1.4.1 Damped Oscillations

Consider a system of mass m attached to a spring of spring constant k and subject to a resistive force of type $\vec{f} = -b\vec{v} = -b\dot{x}\hat{x}$

$$m\ddot{x} + b\dot{x} + kx = 0$$

The constant $\dfrac{b}{m} = 2\beta$ is called the damping constant, while $\omega_0 = \sqrt{\dfrac{k}{m}}$ is the natural frequency – the frequency the system would oscillate without the resistive force. The differential equation can be written as

$$\ddot{x} + 2\beta\dot{x} + \omega_0^2 x = 0$$

with the auxiliary equation

$$r^2 + 2\beta r + \omega_0^2 = 0$$

and the solution

$$x(t) = e^{-\beta t}\left(C_1 e^{\sqrt{\beta^2 - \omega_0^2}\, t} + C_2 e^{-\sqrt{\beta^2 - \omega_0^2}\, t} \right)$$

5.1.4.2   Weak Damping $\beta < \omega_0$

Frequency (smaller than the natural frequency) $\omega_1 = \sqrt{\omega_0^2 - \beta^2}$
   has the solution of the form

$$x(t) = Ae^{-\beta t}\cos(\omega_1 t - \delta)$$

5.1.4.3   Critical Damping $\beta = \omega_0$

The solution has the form

$$x(t) = C_1 e^{-\beta t} + C_2\, t\, e^{-\beta t}$$

5.1.4.4   Strong Damping $\beta > \omega_0$

The solution has the form

$$x(t) = C_1 e^{-\left(\beta - \sqrt{\beta^2 - \omega_0^2}\right)t} + C_2 e^{-\left(\beta + \sqrt{\beta^2 - \omega_0^2}\right)t}$$

5.1.4.5   Driven Damped Oscillations

In the case of driven damped oscillators, besides Hooke's law force and the resistive force, now there is another external force which supports the oscillation, called the driving force $F(t)$.

$$m\ddot{x} + b\dot{x} + kx = F(t)$$

With $f(t) = \dfrac{F(t)}{m}$, the equation of motion is written as

$$\ddot{x} + 2\beta\dot{x} + \omega_0^2 x = f(t)$$

Usually, the driving force is of the sine or cosine type.

5.2   PROBLEMS AND SOLUTIONS

PROBLEM 5.1
A cat is playing with a toy connected to a spring fixed to the wall on the other end. The toy has a mass of $m = 50\,\text{g}$ and the spring constant is $k = 20\,\text{N/m}$ and is following a simple harmonic motion with the expression

$$x(t) = A\cos(\omega t - \delta)$$

a. Calculate the angular frequency ω, the frequency f, and the period τ.
b. If the cat stretches the spring by an initial length $x_0 = 10\,\text{cm}$, and keeps the toy at rest for an instant, identify the amplitude A of the motion and calculate the phase δ.
c. In another playful moment, the toy has the speed of 4 m/s at the equilibrium position of the spring $x_0 = 0$. What are the amplitude A and the phase δ in this case?

SOLUTION 5.1

a. The angular frequency is related to the spring constant k and the mass m by the following relationship:

$$\omega^2 = \frac{k}{m}$$

$$\omega = \sqrt{\frac{k}{m}} = \sqrt{\frac{20\,\dfrac{N}{m}}{0.05\,\text{kg}}} = 20\,\text{s}^{-1}$$

The frequency is obtained from the equation

$$\omega = 2\pi f$$

$$f = \frac{\omega}{2\pi} = \frac{20\,\text{s}^{-1}}{2\pi} = 3.18\,\text{Hz}$$

The period is immediately following, as the inverse of the frequency

$$\tau = \frac{1}{f} = 0.32\,s$$

b. The simple harmonic motion has the expression $x(t) = A\cos(\omega t - \delta)$. The velocity is

$$v(t) = \frac{dx(t)}{dt} = \dot{x}(t) = -\omega A\sin(\omega t - \delta)$$

The initial conditions are $x_0 = x(0) = 10\,\text{cm}$ and $v_0 = v(0) = 0$.

In the initial expression, for the simple harmonic motion, and with using the properties of the trigonometrical functions $\cos(-\delta) = \cos\delta$, and $\sin(-\delta) = -\sin\delta$,

$$x(t = 0) = A\cos(-\delta) = A\cos\delta$$

The phase can be found from the second condition,

$$v_0 = 0$$

$$v(t = 0) = -\omega A\sin(-\delta) = \omega A\sin\delta = 0$$

$$\delta = 0$$

Back to the first initial condition,

$$x(t = 0) = A\cos\delta = A\cos 0 = A = 10\,\text{cm} = 0.1\,\text{m}$$

c. In a similar way, starting with the expression for the position and velocity, and, considering the initial condition, the amplitude and phase are calculated as

$$x(t) = A\cos(\omega t - \delta)$$

and

$$v(t) = -\omega A\sin(\omega t - \delta)$$

So

$$x_0 = 0$$

$$x(t = 0) = A\cos(-\delta) = A\cos\delta = 0$$

$$\delta = \frac{\pi}{2}$$

And from

$$v_0 = 4\,\frac{m}{s}$$

$$v(t = 0) = -\omega A\sin(-\delta) = \omega A\sin\delta = \omega A\sin\frac{\pi}{2} = \omega A$$

$$A = \frac{v_0}{\omega} = \frac{4\,\dfrac{m}{s}}{20\,\dfrac{1}{s}} = 0.2\,\text{m}$$

PROBLEM 5.2

a. An oscillation has the solution

$$x(t) = \beta_1 \cos(\omega t) + \beta_2 \sin(\omega t)$$

Given initial conditions $x(0) = 0$ and $v(0) = 2$, solve for β_1 and β_2.

b. An oscillation has the solution

$$x(t) = A\cos(\omega t - \delta)$$

Given initial conditions $x(0) = 3$ and $v(0) = 1$, solve for A and δ.

SOLUTION 5.2

a. Since $\sin(0) = 0$ and $\cos(0) = 1$, $x(0) = \beta_1 = 0$. The equation can be simplified to

$$x(t) = \beta_2 \sin(\omega t)$$

$v(t) = \dot{x}(t)$ is obtained by differentiation

$$\dot{x}(t) = \omega \beta_2 \cos(\omega t)$$

Thus, $v(0) = \omega \beta_2 = 2$. Solving for β_2, the result is

$$\beta_2 = \frac{2}{\omega}$$

The solution of the oscillation is $x(t) = \dfrac{2}{\omega}\sin(\omega t)$.

b. The derivative of $x(t)$ is

$$\dot{x}(t) = v(t) = -\omega A \sin(\omega t - \delta)$$

From the given initial conditions, a system of two equations is obtained:

$$\begin{cases} A\cos(-\delta) = 3 \\ \omega A\sin(-\delta) = 1 \end{cases}$$

After solving the system, values for δ and A are obtained:

$$\delta = \arctan\left(-\frac{1}{3\omega}\right)$$

$$A = \sqrt{9 + \left(\frac{1}{\omega}\right)^2}$$

PROBLEM 5.3

An object of mass $m = 0.1\,\text{kg}$ is in a simple oscillatory motion defined by equation of motion $x(t) = 5\cos\left(3\pi t - \dfrac{\pi}{4}\right)$, in centimeters.

a. Find the amplitude, the angular frequency, the phase δ, the period τ, and the frequency f.
b. Calculate the spring constant.
c. What are the position and the velocity at $t = 0\,\text{s}$?
d. Find the acceleration at $t = 0\,\text{s}$.

SOLUTION 5.3

a. The general solution in simple harmonic motion is

$$x(t) = A\cos(\omega t - \delta)$$

The given equation is

$$x(t) = 5\cos\left(3\pi t - \frac{\pi}{4}\right)$$

By identification, $A = 5\,\text{cm}$, $\omega = 3\pi\,\text{s}^{-1}$, and $\delta = \dfrac{\pi}{4}$. From here, frequency and period can be easily calculated as follows:

$$\omega = 2\pi f$$

$$f = \frac{\omega}{2\pi} = \frac{3\pi\,\text{s}^{-1}}{2\pi} = 1.5\,\text{Hz}$$

$$\tau = \frac{1}{f} = \frac{1}{1.5\,\text{Hz}} = 0.67\,\text{s}$$

b. The spring constant can be easily calculated from

$$\omega^2 = \frac{k}{m}$$

$$k = m\omega^2 = 0.1\,\text{kg}\left(3\pi\,\text{s}^{-1}\right)^2 = 8.88\,\frac{\text{N}}{\text{m}}$$

c. The initial position is

$$x_0 = x(t = 0) = 5\cos\left(-\frac{\pi}{4}\right) = 5\cos\left(\frac{\pi}{4}\right) = \frac{5\sqrt{2}}{2}\,\text{cm} = 3.54\,\text{cm}$$

The velocity is

$$v(t) = -\omega A \sin(\omega t - \delta)$$

$$v_0 = v(t = 0) = -\omega A \sin(-\delta) = \omega A \sin \delta = 15\pi \sin\frac{\pi}{4}\frac{cm}{s}$$

$$= 15\pi \frac{\sqrt{2}}{2}\frac{cm}{s} = 33.3\frac{cm}{s} = 3.33 \times 10^{-1}\,m/s$$

d. The acceleration is obtained by differentiating the velocity with respect to time

$$a(t) = \frac{dv(t)}{dt} = -\omega^2 A \cos(\omega t - \delta) = -\omega^2 x(t)$$

The initial acceleration can be easily calculated as

$$a_0 = a(t = 0) = -\omega^2 A \cos(-\delta) = -(3\pi)^2 5 \cos\left(-\frac{\pi}{4}\right)\frac{cm}{s^2} = -(3\pi)^2 5 \cos\left(\frac{\pi}{4}\right)\frac{cm}{s^2}$$

$$= -45\pi^2 \frac{\sqrt{2}}{2}\frac{cm}{s^2} = -314\frac{cm}{s^2} = -3.14\,m/s^2$$

PROBLEM 5.4
An underdamped oscillation has sinusoidal frequency of $\sqrt{\omega^2 - \beta^2}$. Consider an underdamped oscillator

$$x(t) = A e^{-\beta t} \cos\left(t\sqrt{\omega^2 - \beta^2}\right)$$

a. Given $\beta = \frac{\omega}{5}$, at what time t_h is the amplitude of the oscillation half of the initial amplitude?

b. The period of oscillation is $\dfrac{2\pi}{\sqrt{\omega^2 - \beta^2}}$. What is the amplitude after one period, given $\beta = \dfrac{\omega}{a}$, with $a > 1$.

SOLUTION 5.4

a. The amplitude is given by $A e^{-\beta t}$. Thus, the equation $A e^{-\beta t_h} = \frac{1}{2}A$ needs to be solved for t_h.

$$A e^{-\beta t_h} = \frac{1}{2}A$$

$$e^{-\frac{\omega}{5}t_h} = \frac{1}{2}$$

$$t_h = \frac{5\ln 2}{\omega}$$

Therefore, at time $t_h = \dfrac{5\ln 2}{\omega}$, the amplitude of the oscillation is half of the original amplitude.

b. The amplitude A_p at $t = \dfrac{2\pi}{\sqrt{\omega^2 - \beta^2}}$ is the amplitude of oscillation after one period.

$$A_p = Ae^{-\frac{\omega}{a} \cdot \frac{2\pi}{\sqrt{\omega^2 - \frac{\omega^2}{a^2}}}}$$

$$A_p = Ae^{-\frac{\omega}{a} \cdot \frac{2\pi}{\sqrt{\omega^2\left(1 - \frac{1}{a^2}\right)}}}$$

$$A_p = Ae^{-\frac{2\pi}{\sqrt{a^2 - 1}}}$$

The amplitude of the oscillation after one period is

$$A_p = Ae^{-\frac{2\pi}{\sqrt{a^2 - 1}}}$$

PROBLEM 5.5
A critically damped oscillator, where $\beta = \omega = \sqrt{\dfrac{k}{m}}$, has the solution

$$x(t) = C_1 e^{-\beta t} + C_2 t e^{-\beta t}$$

a. Check that $x(t)$ satisfies the differential equation $\ddot{x} + 2\beta\dot{x} + \omega^2 x = 0$.
b. Given $x(0) = 2$ and $\dot{x}(0) = 1$, solve for C_1 and C_2.

SOLUTION 5.5

a. The first derivative of x is

$$\dot{x}(t) = -C_1\beta e^{-\beta t} - C_2\beta t e^{-\beta t} + C_2 e^{-\beta t}$$

$$\dot{x}(t) = e^{-\beta t}(-C_1\beta + C_2) - C_2\beta t e^{-\beta t}$$

The second derivative of x is

$$\ddot{x}(t) = C_1\beta^2 e^{-\beta t} + C_2\beta^2 t e^{-\beta t} - 2C_2\beta e^{-\beta t}$$

$$\ddot{x}(t) = e^{-\beta t}(C_1\beta^2 - 2C_2\beta) + C_2\beta^2 t e^{-\beta t}$$

Using the derivatives, it follows that

$$\ddot{x} + 2\beta\dot{x} + \omega^2 x = e^{-\beta t}(C_1\beta^2 - 2C_2\beta) + C_2\beta^2 t e^{-\beta t} + e^{-\beta t}(-2C_1\beta^2 + 2C_2)$$
$$-2C_2\beta^2 t e^{-\beta t} + C_1\beta^2 e^{-\beta t} + C_2\beta^2 t e^{-\beta t} = 0$$

So $x(t)$ satisfies $\ddot{x} + 2\beta\dot{x} + \omega^2 x = 0$.

b. Using $x(0) = 2$, the value for $C_1 = 2$. The first derivative of x is

$$\dot{x}(t) = e^{-\beta t}(-C_1\beta + C_2) - C_2\beta t e^{-\beta t}$$

Since $\dot{x}(0) = 1$,

$$-C_1\beta + C_2 = 1$$

Thus, $C_2 = 2\beta + 1$. Therefore, the solution to the oscillation is

$$x(t) = 2e^{-\beta t} + (2\beta + 1)t e^{-\beta t}$$

PROBLEM 5.6
Consider the case of an overdamped oscillation where $\beta = a\omega$, with $a > 1$, with solution

$$x(t) = e^{-\left(\beta - \sqrt{\beta^2 - \omega^2}\right)t} + e^{-\left(\beta + \sqrt{\beta^2 - \omega^2}\right)t}$$

a. What is the value of $x\left(\dfrac{2\pi}{\omega}\right)$?

b. Find $x\left(\dfrac{2\pi}{\omega}\right)$ for $a = 2, 5$, and 10.

SOLUTION 5.6
a. First, the equation of motion of the oscillation can be rewritten using the relation $\beta = a\omega$.

$$x(t) = e^{-\left(a\omega - \sqrt{a^2\omega^2 - \omega^2}\right)t} + e^{-\left(a\omega + \sqrt{a^2\omega^2 - \omega^2}\right)t}$$

$$x(t) = e^{-\left(a\omega - \sqrt{\omega^2\left(a^2-1\right)}\right)t} + e^{-\left(a\omega + \sqrt{\omega^2\left(a^2-1\right)}\right)t}$$

$$x(t) = e^{-\omega\left(a - \sqrt{a^2-1}\right)t} + e^{-\omega\left(a + \sqrt{a^2-1}\right)t}$$

Now, the value $x\left(\dfrac{2\pi}{\omega}\right)$ can be computed:

$$x\left(\frac{2\pi}{\omega}\right) = e^{-\omega\left(a-\sqrt{a^2-1}\right)\frac{2\pi}{\omega}} + e^{-\omega\left(a+\sqrt{a^2-1}\right)\frac{2\pi}{\omega}}$$

$$x\left(\frac{2\pi}{\omega}\right) = e^{-2\pi\left(a-\sqrt{a^2-1}\right)} + e^{-2\pi\left(a+\sqrt{a^2-1}\right)}$$

b. For $a = 2$,

$$x\left(\frac{2\pi}{\omega}\right) = e^{-2\pi\left(2-\sqrt{2^2-1}\right)} + e^{-2\pi\left(2+\sqrt{2^2-1}\right)} = 0.1857$$

For $a = 5$,

$$x\left(\frac{2\pi}{\omega}\right) = e^{-2\pi\left(5-\sqrt{5^2-1}\right)} + e^{-2\pi\left(5+\sqrt{5^2-1}\right)} = 0.5301$$

For $a = 10$,

$$x\left(\frac{2\pi}{\omega}\right) = e^{-2\pi\left(10-\sqrt{10^2-1}\right)} + e^{-2\pi\left(10+\sqrt{10^2-1}\right)} = 0.7298$$

PROBLEM 5.7

A toy car is connected to a spring and moving as a critically damped oscillator. The general solution for a critically damped oscillator is $x(t) = e^{-\omega_0 t}(C_1 + C_2 t)$, where ω_0 is the natural frequency in the absence of any resistive forces.

a. The toy is pulled at a position x_0 from the equilibrium position, then the toy is released from rest. Find the equation of motion (position vs. time) and the velocity versus time.

b. This time the system spring car is passing through the equilibrium position $x_0 = 0$ with a speed v_0. In this case, find the equation of motion $x(t)$ and the velocity $v(t)$.

SOLUTION 5.7

a. The general solution for a critically damped oscillator is

$$x(t) = e^{-\omega_0 t}(C_1 + C_2 t)$$

where C_1 and C_2 are to be determined from the initial conditions.

$$x_0 = x(t = 0) = e^0(C_1 + C_2 \cdot 0) = C_1$$

$$C_1 = x_0$$

The velocity is obtained by,

$$v(t) = \frac{dx(t)}{dt} = \dot{x}(t) = -\omega_0 e^{-\omega_0 t}(C_1 + C_2 t) + C_2 e^{-\omega_0 t} = e^{-\omega_0 t}(-\omega_0 C_1 - \omega_0 C_2 t + C_2)$$

The initial velocity is, considering that $C_1 = x_0$

$$v_0 = v(t = 0) = -\omega_0 C_1 + C_2 = -\omega_0 x_0 + C_2$$

Since $v_0 = 0$, it yields that $C_2 = \omega_0 x_0$. With the constants calculated, the equation of motion and the velocity become

$$x(t) = e^{-\omega_0 t}(x_0 + \omega_0 x_0 t) = x_0 e^{-\omega_0 t}(1 + \omega_0 t)$$

$$v(t) = e^{-\omega_0 t}(-\omega_0 x_0 - \omega_0^2 x_0 t + \omega_0 x_0) = -\omega_0^2 x_0 t e^{-\omega_0 t}$$

b. General solution for a critically damped oscillator is

$$x(t) = e^{-\omega_0 t}(C_1 + C_2 t)$$

In this case, the initial conditions are $x_0 = 0$ and v_0

$$x_0 = x(t = 0) = C_1$$

$$C_1 = 0$$

The velocity is, as calculated before,

$$v(t) = \frac{dx(t)}{dt} = e^{-\omega_0 t}(-\omega_0 C_1 - \omega_0 C_2 t + C_2)$$

$$v_0 = v(t = 0) = -\omega_0 C_1 + C_2 = C_2$$

$$C_2 = v_0$$

Consequently, the equation of motion and the velocity are, in this case,

$$x(t) = v_0 t \, e^{-\omega_0 t}$$

$$v(t) = v_0(1 - \omega_0 t)e^{-\omega_0 t}$$

PROBLEM 5.8

In a simple harmonic motion

$$x(t) = A\cos(\omega t - \delta)$$

the motion is fully characterized by the amplitude A, the angular frequency ω, and the phase δ.

If the amplitude and angular frequency are not known, they can be found by analyzing the motion and measuring the position and the velocity of the oscillator at two different moments.

Find A and ω in two different ways.

SOLUTION 5.8

a) First method, by using trigonometry, specifically the equation $(\sin \alpha)^2 + (\cos \alpha)^2 = 1$.
The position and the velocity are

$$x(t) = A\cos(\omega t - \delta)$$

and

$$v(t) = -\omega A \sin(\omega t - \delta)$$

For two different moments t_1 and t_2, the position and the velocities are

$$x_1 = A\cos(\omega t_1 - \delta)$$
$$v_1 = -\omega A \sin(\omega t_1 - \delta)$$

and

$$x_2 = A\cos(\omega t_2 - \delta)$$
$$v_2 = -\omega A \sin(\omega t_2 - \delta)$$

By substituting sine and cosine from each moment and adding the squares,

$$\frac{x_1^2}{A^2} + \frac{v_1^2}{\omega^2 A^2} = 1$$

$$\frac{1}{A^2}\left(x_1^2 + \frac{v_1^2}{\omega^2}\right) = 1$$

$$A^2 = x_1^2 + \frac{v_1^2}{\omega^2}$$

Similarly,

$$\frac{x_2^2}{A^2} + \frac{v_2^2}{\omega^2 A^2} = 1$$

$$A^2 = x_2^2 + \frac{v_2^2}{\omega^2}$$

From both equations, by transitivity,

$$x_1^2 + \frac{v_1^2}{\omega^2} = x_2^2 + \frac{v_2^2}{\omega^2}$$

$$x_1^2 - x_2^2 = \frac{v_2^2 - v_1^2}{\omega^2}$$

$$\omega^2 = \frac{v_2^2 - v_1^2}{x_1^2 - x_2^2}$$

From here, substituting in one of the equations,

$$A^2 = x_1^2 + \frac{v_1^2}{\omega^2} = x_1^2 + v_1^2 \frac{x_1^2 - x_2^2}{v_2^2 - v_1^2} = \frac{x_1^2 v_2^2 - x_1^2 v_1^2 + x_1^2 v_1^2 - x_2^2 v_1^2}{v_2^2 - v_1^2} = \frac{x_1^2 v_2^2 - x_2^2 v_1^2}{v_2^2 - v_1^2}$$

b) Second method, using the conservation of total mechanical energy in the first state, at time t_1, characterized by position x_1 and velocity v_1 and state 2 at time t_2, characterized by position x_2 and velocity v_2

$$E = E_1 = T_1 + U_1 = \frac{mv_1^2}{2} + \frac{kx_1^2}{2} = E_2 = T_2 + U_2 = \frac{mv_2^2}{2} + \frac{kx_2^2}{2}$$

$$m(v_1^2 - v_2^2) = k(x_2^2 - x_1^2)$$

Recalling that $\omega^2 = \dfrac{k}{m}$ it follows that

$$\omega^2 = \frac{v_1^2 - v_2^2}{x_2^2 - x_1^2} = \frac{v_2^2 - v_1^2}{x_1^2 - x_2^2}$$

Also, when $x(t) = A$ the potential energy is maximum, and the kinetic energy is zero.

$$E = E_1 = E_2 = \frac{kA^2}{2} = \frac{mv_1^2}{2} + \frac{kx_1^2}{2}$$

$$A^2 = \frac{m}{k}v_1^2 + x_1^2 = \frac{v_1^2}{\omega^2} + x_1^2 = \frac{v_1^2(x_2^2 - x_1^2)}{v_1^2 - v_2^2} + x_1^2$$

$$= \frac{v_1^2 x_2^2 - v_1^2 x_1^2 + v_1^2 x_1^2 - v_2^2 x_1^2}{v_1^2 - v_2^2} = \frac{v_1^2 x_2^2 - v_2^2 x_1^2}{v_1^2 - v_2^2} = \frac{x_1^2 v_2^2 - x_2^2 v_1^2}{v_2^2 - v_1^2}$$

Both methods lead to the same results. The energy method is more physical, based on energies. However, the first method can also be employed.

PROBLEM 5.9

Consider a block of mass m connected to a spring with a constant k on a ramp inclined at an angle θ (Figure 5.1). Find an expression for the spring length if the spring's rest length is d.

SOLUTION 5.9

Since the block is stationary, Newton's Second Law is given by

$$\Sigma F = 0$$

Considering the following free body diagram (Figure 5.2).
It is clear the gravitational component pulling on the spring is

$$F_g = mg \sin\theta$$

The spring length is thus given by $L = d + x_0$ where x_0 is the amount the spring stretches. Substituting everything into Newton's Second Law yields

$$mg \sin\theta - kx_0 = 0$$

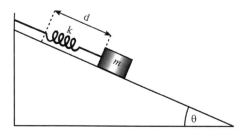

FIGURE 5.1 Block on a ramp connected to a spring.

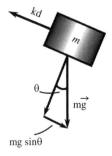

FIGURE 5.2   Free body diagram of the block.

$$x_0 = \frac{mg}{k} \sin \theta$$

Therefore, the spring's length is given by

$$L = d + \frac{mg}{k} \sin \theta$$

PROBLEM 5.10

Consider the mass m connected to two springs above and below with spring constants k_1 and k_2, respectively, as in Figure 5.3. Find the angular velocity ω after a small displacement in the y direction.

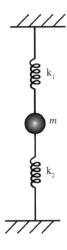

FIGURE 5.3   Mass connected to a spring above and a spring below.

SOLUTION 5.10

To find the mass's acceleration, consider Newton's Second Law

$$\sum F = m\ddot{y}$$

Considering the mass is displaced by an amount y downward, this becomes

$$k_1 y + k_2 y - mg = m\ddot{y}$$

$$\ddot{y} = \frac{k_1 + k_2}{m} y - g$$

Noting that $y < 0$

$$\ddot{y} = -\frac{k_1 + k_2}{m} |y| - g$$

and therefore, the angular velocity is

$$\omega = \frac{k_1 + k_2}{m}$$

It is important to note that assuming the displacement was downward does not impact the solution. If instead the displacement was upward, Newton's Second Law yields

$$-k_1 y - k_2 y - mg = m\ddot{y}$$

$$\ddot{y} = -\frac{k_1 + k_2}{m} y - g$$

Since $y > 0$ here, the absolute value of y is not necessary and the angular velocity is exactly what was found assuming the opposite displacement.

PROBLEM 5.11

Consider the mass–spring system in Figure 5.4 consisting of two springs with constants k and a mass m. When the mass is in line with the springs, they are at their rest length l_0. The mass is then gently allowed to descend (to prevent oscillations) until

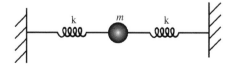

FIGURE 5.4   Mass connected to a spring to the left and a spring to the right.

the system is in equilibrium. Find the amount the springs stretch, assuming the final position is a small angle away from the initial position.

SOLUTION 5.11

After the mass is allowed to come to rest, the system is in the position as shown in Figure 5.5.

The amount the springs stretch can be expressed as Δl, which is illustrated in Figure 5.6.

From this triangle,

$$\cos\theta = \frac{l_0}{\Delta l + l_0}$$

To find an expression for θ, Newton's Second Law can be used, considering the mass is stationary

$$\Sigma F_y = 0$$

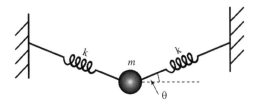

FIGURE 5.5 Mass connected to springs on the left and the right in its final position.

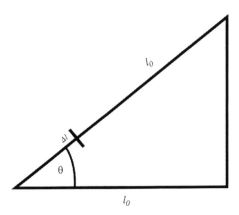

FIGURE 5.6 Triangle illustrating the equilibrium lengths of the system.

Therefore,

$$2k\Delta l \sin\theta - mg = 0$$

where $k\Delta l \sin\theta$ is the amount of force one spring acts on the mass, in the y direction. Assuming small angles, these equations become

$$1 - \frac{\theta^2}{2} = \frac{l_0}{\Delta l + l_0}$$

and

$$2k\Delta l\theta = mg$$

Since Δl is the quantity of interest, the second equation can be rewritten as

$$\theta = \frac{mg}{2k\Delta l}$$

Substituting this into the first equation yields

$$1 - \frac{m^2 g^2}{8k^2(\Delta l)^2} = \frac{l_0}{\Delta l + l_0}$$

$$8k^2(\Delta l)^2 - m^2 g^2 = \frac{8k^2(\Delta l)^2 l_0}{\Delta l + l_0}$$

$$8k^2(\Delta l)^3 + 8k^2(\Delta l)^2 l_0 - m^2 g^2 \Delta l - m^2 g^2 l_0 = 8k^2(\Delta l)^2 l_0$$

$$8k^2(\Delta l)^3 - m^2 g^2 \Delta l - m^2 g^2 l_0 = 0$$

$$(\Delta l)^3 - \frac{m^2 g^2}{8k^2}\Delta l - \frac{m^2 g^2}{8k^2}l_0 = 0$$

This is a cubic which has the solution

$$\Delta l = \sqrt[3]{\frac{m^2 g^2 l_0}{16k^2} + \sqrt{\left(\frac{m^2 g^2 l_0}{16k^2}\right)^2 + \left(-\frac{m^2 g^2}{24k^2}\right)^3}}$$

$$+ \sqrt[3]{\frac{m^2 g^2 l_0}{16k^2} - \sqrt{\left(\frac{m^2 g^2 l_0}{16k^2}\right)^2 + \left(-\frac{m^2 g^2}{24k^2}\right)^3}}$$

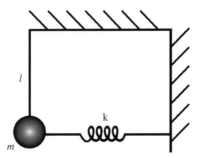

FIGURE 5.7   Mass connected to a pendulum and a spring.

PROBLEM 5.12

Consider a mass m attached to a pendulum of length l and a spring of constant k as depicted in Figure 5.7. When the mass is at the lowest point of the pendulum, the spring is at its rest length. Considering a small displacement, find the angular velocity.

SOLUTION 5.12

When the mass is displaced, gravity is pulling the mass down toward the equilibrium pendulum position and the spring is pushing/pulling it toward its equilibrium position. If the mass is displaced by an angle ϕ, the following restoring forces are at play

$$F_\phi = mg \sin \phi$$

$$F_s = kl \tan \phi$$

Considering small angles, the $\hat{\phi}$ forces due to gravity and the spring are approximately in the same direction. The forces also become

$$mg \sin \phi \rightarrow mg\phi$$

$$kl \tan \phi \rightarrow kl\phi$$

Newton's Second Law is then given by

$$\Sigma F_\phi = ml\ddot{\phi}$$

$$-kl\phi - mg\phi = ml\ddot{\phi}$$

$$\ddot{\phi} = -\frac{kl + mg}{ml}\phi$$

and the angular velocity is

$$\omega^2 = \frac{kl + mg}{ml}$$

$$\omega = \sqrt{\frac{kl + mg}{ml}}$$

PROBLEM 5.13

A small cylinder of mass m is sliding without friction along a thin fixed rod (Figure 5.8). If the spring of spring constant k is elongated a distance h by the force F_0, find the frequency of the oscillations performed by the cylinder. Consider the case of small oscillations.

SOLUTION 5.13

The spring elongation is

$$\delta = \sqrt{h^2 + x^2} - h$$

For $h \gg x$

$$\delta = h\sqrt{1 + \left(\frac{x}{h}\right)^2} - h = h\left(1 + \frac{x^2}{2h^2} - 1\right) = \frac{x^2}{2h}$$

Potential energy can be approximated with

$$U = F_0\delta = F_0 \frac{x^2}{2h}$$

On another side, $F_0 = kh$ and recalling that $\omega^2 = \dfrac{k}{m}$ and substituting k, it yields that the frequency of the small oscillations is

$$\omega = \sqrt{\frac{F_0}{mh}}$$

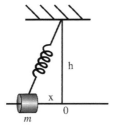

FIGURE 5.8 Small cylinder of mass m oscillating on a thin rod.

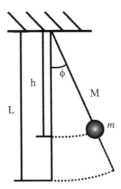

FIGURE 5.9 Mass m changing position on a rod of mass M fixed on one side and oscillating in a vertical plane.

PROBLEM 5.14

A uniform branch of mass M and length L moves like a simple pendulum in the wind. A ladybug of mass m is climbing the branch and stops to rest in different positions h from the top of the branch (Figure 5.9). Find the angular frequency, the period of the small oscillations, and the value of h for which we obtain the smallest period. What is the minimum period?

SOLUTION 5.14

The kinetic energy of the system branch (mass M) and ladybug (mass m) is

$$T = \frac{I\omega^2}{2} + \frac{mv^2}{2}$$

The moment of inertia I of a thin rod of mass M and length L about the axis through one end perpendicular to the rod is $I = \dfrac{ML^2}{3}$ and the angular velocity $\omega = \dfrac{d\phi}{dt} = \dot{\phi}$. The kinetic energy becomes

$$T = \frac{I\omega^2}{2} + \frac{mv^2}{2} = \frac{\dfrac{ML^2}{3}\dot{\phi}^2}{2} + \frac{mh^2\dot{\phi}^2}{2} = \frac{1}{2}\left(\frac{ML^2}{3} + mh^2\right)\dot{\phi}^2$$

The potential energy is, considering that the center of the mass for the uniform rod is at mid-distance,

$$U = \frac{M}{2}gL(1-\cos\phi) + mgh(1-\cos\phi) = \left(\frac{ML}{2} + mh\right)g(1-\cos\phi)$$

Equilibrium is obtained by deriving the potential energy by the variable ϕ

$$\frac{dU}{d\phi} = \left(\frac{ML}{2} + mh\right)g\sin\phi$$

Equilibrium condition $\dfrac{dU}{d\phi} = 0$ leads to $\sin\phi = 0$, therefore $\phi = 0$ is the equilibrium position.

Stable equilibrium near $\phi = 0$, and using the approximation of the small oscillations, $\cos\phi \cong 1 - \dfrac{\phi^2}{2}$

$$U = \left(\frac{ML}{2} + mh\right)g(1 - \cos\phi) \cong \left(\frac{ML}{2} + mh\right)g\left(1 - 1 + \frac{\phi^2}{2}\right) = \frac{1}{2}\left(\frac{ML}{2} + mh\right)g\phi^2$$

$$\frac{d^2U}{d\phi^2} = \left(\frac{ML}{2} + mh\right)g\cos\phi \cong \left(\frac{ML}{2} + mh\right)g\left(1 - \frac{\phi^2}{2} + \cdots\right)$$

Or, keeping just the first term, $\dfrac{d^2U}{d\phi^2} = \left(\dfrac{ML}{2} + mh\right)g$.

Back to the kinetic energy,

$$T = \frac{1}{2}\left(\frac{ML^2}{3} + mh^2\right)\dot{\phi}^2 = \frac{1}{2}f\dot{\phi}^2$$

where

$$f = \left(\frac{ML^2}{3} + mh^2\right)$$

So

$$\omega^2 = \frac{1}{f}\left(\frac{d^2U}{d\phi^2}\right)_{\phi=0}$$

$$\omega^2 = \frac{\left(\dfrac{ML}{2} + mh\right)g}{\left(\dfrac{ML^2}{3} + mh^2\right)} = \frac{ML + 2mh}{ML^2 + 3mh^2}\frac{3g}{2}$$

$$\omega = \sqrt{\frac{ML + 2mh}{ML^2 + 3mh^2} \frac{3g}{2}}$$

The period of small oscillations is τ with $\omega = \dfrac{2\pi}{\tau}$ and

$$\tau = \frac{2\pi}{\omega} = 2\pi \sqrt{\frac{ML^2 + 3mh^2}{ML + 2mh} \frac{2}{3g}}$$

The period is minimum for

$$\frac{d\tau}{dh} = 0$$

$$\frac{d\tau}{dh} = 2\pi \sqrt{\frac{2}{3g}} \frac{1}{2} \left(\frac{ML^2 + 3mh^2}{ML + 2mh} \right)^{-\frac{1}{2}} \frac{\left[6mh(ML + 2mh) - (ML^2 + 3mh^2)2m \right]}{(ML + 2mh)^2}$$

$$= \pi \sqrt{\frac{2}{3g}} \sqrt{\frac{ML + 2mh}{ML^2 + 3mh^2}} \frac{2m(3hML + 6mh^2 - ML^2 - 3mh^2)}{(ML + 2mh)^2}$$

$$= \pi \sqrt{\frac{2}{3g}} \sqrt{\frac{1}{ML^2 + 3mh^2}} \frac{2m(3hML + 3mh^2 - ML^2)}{(ML + 2mh)^{1\frac{1}{2}}}$$

The condition for minimum period is

$$\frac{d\tau}{dh} = 0$$

$$3mh^2 + 3hML - ML^2 = 0$$

After dividing by $3m$, the equation is

$$h^2 + h\frac{M}{m}L - \frac{M}{3m}L^2 = 0$$

From here, the minimum length (choosing the positive solution) is

$$h_{min} = \frac{-\dfrac{M}{m}L + \sqrt{\dfrac{M^2}{m^2}L^2 + \dfrac{4ML^2}{3m}}}{2} = \frac{ML}{2m} \left(\sqrt{\frac{4m}{3M} + 1} - 1 \right)$$

The period of small oscillations for the minimum height h_{min} is

$$\tau_{min} = \frac{2\pi}{\omega} = 2\pi\sqrt{\frac{ML^2 + 3mh_{min}^2}{ML + 2mh_{min}}\frac{2}{3g}}$$

And after extensive calculations,

$$\tau_{min} = 2\pi\sqrt{\left(\sqrt{\frac{4m}{3M}+1}-1\right)\frac{ML}{2mg}} \cong 2\pi\sqrt{\frac{h_{min}}{g}}$$

PROBLEM 5.15

In damped one-dimensional oscillations, the object of mass m is moving under the influence of Hooke's law force $-kx$ and under the influence of the resistive force

$$f = -bv = -b\dot{x}$$

leading to Newton's Second Law equation of the following form:

$$m\ddot{x} + b\dot{x} + kx = 0$$

Considering an RLC (resistor–inductor–capacitor) circuit in series in Figure 5.10, write Kirchhoff's loop equation and, by comparison, find which quantity plays the role of the mass, of the resistive constant b, and of spring constant k. Also, find the resonance frequency ω_0.

SOLUTION 5.15

The circuit does not have a battery, so the total voltage on the resistor, on the capacitor, and on the inductance should be zero, $V_R + V_L + V_C = 0$.

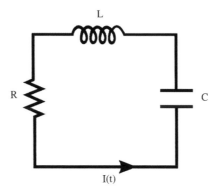

FIGURE 5.10 A resistor–inductor–capacitor (RLC) circuit in series with no power supply.

Recalling that the electric current is the derivative of electric charge with respect to time,

$$I(t) = \frac{dq(t)}{dt} = \dot{q}(t), V_R = RI = R\dot{q}$$

The voltage on the inductor is

$$V_L = L\frac{dI(t)}{dt} = L\frac{d^2q(t)}{dt^2} = L\ddot{q}$$

And the voltage on the capacitor is

$$V_L = \frac{q(t)}{C} = \frac{q}{C}$$

Finally, the sum becomes

$$R\dot{q} + L\ddot{q} + \frac{q}{C} = 0$$

or, rearranged

$$L\ddot{q} + R\dot{q} + \frac{q}{C} = 0$$

By comparison with the damped oscillator $m\ddot{x} + b\dot{x} + kx = 0$, we can identify in the role of mass is the impedance L, in the role of the resistive coefficient b, the resistance R, and in the role of spring constant, the inverse of the capacitance C. In mechanics $\frac{b}{m} = 2\beta$, where β is the damping constant, in RLC circuit $\frac{R}{L} = 2\beta$.

Recalling that the natural frequency is $\omega_0 = \frac{k}{m}$. It follows that for the RLC circuit, by equivalence, $\omega_0 = \sqrt{\frac{1}{LC}}$, which is the resonance frequency, in the case when the capacitive reactance is equal to the inductive reactance

$$X_L = X_C$$

$$\omega_0 L = \frac{1}{\omega_0 C}$$

$$\omega_0^2 = \frac{1}{LC}$$

which is indeed the resonance frequency.

PROBLEM 5.16

Suppose there exists a spring network of n springs – this looks very similar to networks of resistors and/or capacitors. Considering this, perhaps there is an analogy between springs and circuit elements. Taking this n springs in parallel (Figure 5.11) and in series (Figure 5.12), determine if such an analogy exists. Explain why, intuitively, this makes sense. Assume all springs have the same constant k and rest length l.

SOLUTION 5.16

Taking the n springs in parallel first, consider the force required to displace the end of all springs is

$$F = \underbrace{kx + kx + \cdots + kx}_{n} = (nk)x = Kx$$

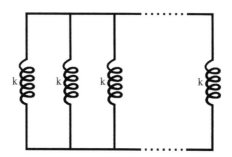

FIGURE 5.11   Network of parallel springs.

FIGURE 5.12   Network of series springs.

Therefore, in parallel, an "equivalent spring" would have constant $K = nk$. Considering the series case next, the amount of force required to displace the end of the last sprint by x requires all the springs in the chain to extend $\frac{x}{n}$. The force is then the amount the last spring in the chain pulls back, specifically

$$F = k\frac{x}{n}$$

In an attempt to relate this back to circuit elements and finding an equivalent spring constant, this can be expressed as

$$\frac{k}{n} = \frac{1}{\frac{n}{k}} = \frac{1}{\underbrace{\frac{1}{k} + \frac{1}{k} + \cdots + \frac{1}{k}}_{n}}$$

Therefore, an equivalent spring constant is given by

$$\frac{1}{K} = \underbrace{\frac{1}{k} + \frac{1}{k} + \cdots + \frac{1}{k}}_{n}$$

Considering these two configurations, it appears springs in parallel behave like capacitors in parallel and springs in series behave like capacitors in series. Intuitively, this makes sense; both springs and capacitor "store" energy so when there are connected in similar ways, it makes sense that they have similar "equivalent element" equations.

6 Lagrangian Formalism

6.1 THEORY

This chapters presents the Lagrangian formalism. While this formalism does not appear very useful in solving simple problems, the difficult problems can be solved elegantly in this manner.

6.1.1 THE LAGRANGIAN

The Lagrangian is

$$\mathcal{L} = T - U$$

where T is the kinetic energy of the system and U is the potential energy of the system.

For a one-dimensional system with x as generalized coordinate, the Lagrange equation is

$$\frac{\partial \mathcal{L}}{\partial x} = \frac{d}{dt}\frac{\partial \mathcal{L}}{\partial \dot{x}}$$

6.1.2 HAMILTON'S PRINCIPLE

The actual path followed by a particle between point 1, reached at time t_1, and point 2, reached at time t_2, is such that the action integral

$$S = \int_{t_1}^{t_2} \mathcal{L} dt$$

is stationary when taken along the actual path.

For a holonomic system of n generalized coordinates $q_1, \ldots, q_n$, the Lagrangian is obtained as

$$\mathcal{L}(q_1, \ldots, q_n, \dot{q}_1, \ldots, \dot{q}_n, t) = T - U$$

A holonomic system has n degrees of freedom and it can be fully described by exactly n generalized coordinates $q_1, \ldots, q_n$.

DOI: 10.1201/9781003365709-6

The condition that the action integral S

$$S = \int_{t_1}^{t_2} \mathcal{L}(q_1,\ldots,q_n,\dot{q}_1,\ldots,\dot{q}_n,t)dt$$

is stationary, implies that the n Lagrange–Euler equations hold true,

$$\frac{\partial \mathcal{L}}{dq_1} = \frac{d}{dt}\frac{\partial \mathcal{L}}{d\dot{q}_1} \qquad \frac{\partial \mathcal{L}}{dq_2} = \frac{d}{dt}\frac{\partial \mathcal{L}}{d\dot{q}_2} \qquad \cdots \qquad \frac{\partial \mathcal{L}}{dq_n} = \frac{d}{dt}\frac{\partial \mathcal{L}}{d\dot{q}_n}$$

Or, simply called Lagrange equations, with i from 1 to n,

$$\frac{\partial \mathcal{L}}{dq_i} = \frac{d}{dt}\frac{\partial \mathcal{L}}{d\dot{q}_i}$$

6.2   PROBLEMS AND SOLUTIONS

PROBLEM 6.1
Consider two objects of masses m_1 and m_2 suspended by an inextensible massless string of length l which passes over a massless frictionless pulley of radius R, as in Figure 6.1.

a. Obtain the Lagrangian.
b. Write the Lagrange equations and obtain the acceleration.
c. Compare with the method using Newton's Second Law.

SOLUTION 6.1
a. Consider the distance x of the mass m_1 to the level of the center of the pulley as the generalized coordinate. Since the string is inextensible, of length l, the position y of the mass m_2 depends on x as $y = l - \pi R - x = -x + \text{constant}$.

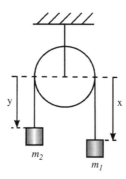

FIGURE 6.1   Atwood machine with two masses.

The speeds of the two objects are also interdependent:

$$\dot{y} = -\dot{x}$$

The kinetic energy is

$$T = \frac{m_1}{2}\dot{x}^2 + \frac{m_2}{2}\dot{y}^2 = \frac{(m_1 + m_2)}{2}\dot{x}^2$$

The potential energy is

$$U = -m_1 gx - m_2 gy = -m_1 gx - m_2 g(-x + \text{const.}) = -(m_1 - m_2)gx + \text{const.}$$

The Lagrangian is

$$\mathcal{L} = T - U = \frac{(m_1 + m_2)}{2}\dot{x}^2 + (m_1 - m_2)gx$$

b. There is only one generalized coordinate for the generalized coordinate x

$$\frac{\partial \mathcal{L}}{\partial x} = \frac{d}{dt}\frac{\partial \mathcal{L}}{\partial \dot{x}}$$

By taking the partial derivatives, it yields that

$$\frac{\partial \mathcal{L}}{\partial x} = (m_1 - m_2)g$$

$$\frac{\partial \mathcal{L}}{\partial \dot{x}} = (m_1 + m_2)\dot{x}$$

And the derivative with respect to time is

$$\frac{d}{dt}\frac{\partial \mathcal{L}}{\partial \dot{x}} = \frac{d}{dt}(m_1 + m_2)\dot{x} = (m_1 + m_2)\ddot{x}$$

The acceleration is easily obtained as expected:

$$\ddot{x} = \frac{(m_1 - m_2)}{(m_1 + m_2)}g$$

Note that the Lagrangian formalism allowed solving the problem without considering the tension in the string, which appears in the introductory physics course when using Newton's Second Law approach.

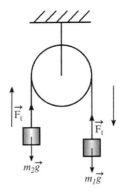

FIGURE 6.2 Atwood machine with two masses and inextensible string, with gravitational forces and the tensions in the string.

c. Using Newton's Second Law applied to each of the two objects (Figure 6.2), and by considering the string inextensible, and the fact that the two objects move with the same magnitude of acceleration, it yields

$$m_1 a = m_1 g - F_t$$

$$m_2 a = -m_2 g + F_t$$

Adding the two equations, the tension in the string F_t is eliminated and the acceleration is obtained as before

$$a = \frac{(m_1 - m_2)}{(m_1 + m_2)} g$$

It is always good to check the solution for particular cases. For $m_1 = m_2$, the acceleration is zero. Note that, while acceleration may be zero for equal masses, the system may move with constant speed if a small nudge is applied to the masses to start moving.

PROBLEM 6.2
Consider two objects of masses m_1 and m_2 suspended by an inextensible massless string of length l which passes over a frictionless pulley of mass M and radius R, as in Figure 6.3.

a. Obtain the Lagrangian.
b. Write the Lagrange equations and obtain the acceleration.

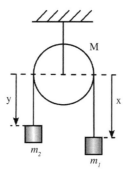

FIGURE 6.3   Atwood machine with two masses and a pulley of mass M.

SOLUTION 6.2

a. The kinetic energy must include the rotational energy due to the pulley of mass M and moment of inertia $I = \dfrac{MR^2}{2}$, and $\omega = \dfrac{v}{R} = \dfrac{\dot{x}}{R}$

$$T = \frac{m_1}{2}\dot{x}^2 + \frac{m_2}{2}\dot{y}^2 + \frac{I\omega^2}{2} = \frac{(m_1+m_2)}{2}\dot{x}^2 + \frac{I\omega^2}{2} = \frac{(m_1+m_2)}{2}\dot{x}^2 + \frac{M\dot{x}^2}{4}$$

The potential energy is the same

$$U = -m_1gx - m_2gy = -m_1gx - m_2g(-x+\text{const.}) = -(m_1-m_2)gx + \text{const.}$$

The Lagrangian is

$$\mathcal{L} = T - U = \frac{\left(m_1 + m_2 + \dfrac{M}{2}\right)}{2}\dot{x}^2 + (m_1-m_2)gx + \text{const}$$

b. There is only one generalized coordinate for the generalized coordinate x

$$\frac{\partial \mathcal{L}}{\partial x} = \frac{d}{dt}\frac{\partial \mathcal{L}}{\partial \dot{x}}$$

By taking the partial derivatives, it yields that

$$\frac{\partial \mathcal{L}}{\partial x} = (m_1-m_2)g$$

$$\frac{\partial \mathcal{L}}{\partial \dot{x}} = \left(m_1 + m_2 + \frac{M}{2}\right)\dot{x}$$

And the derivative with respect to time is

$$\frac{d}{dt}\frac{\partial \mathcal{L}}{\partial \dot{x}} = \frac{d}{dt}\left(m_1 + m_2 + \frac{M}{2}\right)\dot{x} = \left(m_1 + m_2 + \frac{M}{2}\right)\ddot{x}$$

The acceleration is easily obtained as

$$\ddot{x} = \frac{(m_1 - m_2)}{\left(m_1 + m_2 + \dfrac{M}{2}\right)}g$$

Note that the disk of mass M is bringing a term of $\dfrac{M}{2}$ at the denominator, diminishing the acceleration comparing the case with a massless pulley, because some of the energy is used to accelerate the pulley.

PROBLEM 6.3

Consider a block of mass m sliding down a frictionless ramp at an incline θ as in Figure 6.4. Find the velocity of the block at time t if the block is stationary at $t = 0$.

Note: This is the same problem text as Problem 1.4 but now should be solved using Lagrange's equations.

SOLUTION 6.3

Considering the x-axis is parallel to the ramp, the kinetic energy is given by

$$T = \frac{1}{2}m\dot{x}^2$$

Without worrying about the true height of the ramp, the potential energy is given by

$$U = mg(\text{const} - x\sin\theta)$$

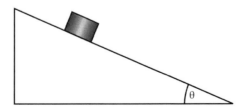

FIGURE 6.4   Block placed on a ramp.

Therefore, the Lagrangian is

$$\mathcal{L} = T - U = \frac{1}{2}m\dot{x}^2 - mg(\text{const} - x\sin\theta)$$

Using Lagrange's equations yields

$$\frac{d}{dt}\frac{\partial\mathcal{L}}{\partial\dot{x}} = \frac{\partial\mathcal{L}}{\partial x}$$

$$\frac{d}{dt}(m\dot{x}) = mg\sin\theta$$

$$\ddot{x} = g\sin\theta$$

The velocity is then given by

$$\dot{x} = gt\sin\theta + C$$

where C is a constant. Since the block is at rest at the top of the ramp,

$$\dot{x}(0) = 0$$

$$C = 0$$

and the velocity as a function of time is

$$\dot{x} = gt\sin\theta$$

which is exactly what was found in previous chapters.

PROBLEM 6.4
A mass m_1 is on an incline attached to a massless string. Another mass m_2 is attached to the other end of the string, hanging vertically, as in Figure 6.5. Find the Lagrangian and obtain the equation of motion.

SOLUTION 6.4
The Lagrangian is equal to the potential energy subtracted from kinetic energy

$$\mathcal{L} = T - U$$

The kinetic energy of the system is

$$T = \frac{1}{2}m_1\dot{x}^2 + \frac{1}{2}m_2\dot{y}^2$$

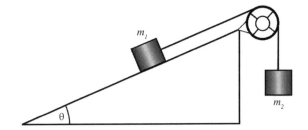

FIGURE 6.5 Masses on incline connected by a string.

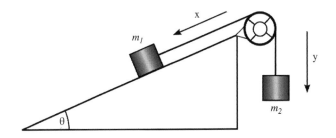

FIGURE 6.6 Masses on incline connected by a string.

The potential energy of the system is

$$U = -m_2 gy - m_1 gx \sin\theta = +m_2 gx - m_1 gx \sin\theta + \text{const}$$

Using these two equations, the Lagrangian is

$$\mathcal{L} = \frac{1}{2}(m_1 + m_2)\dot{x}^2 + (m_1 \sin\theta - m_2)gx + \text{const}$$

Now, the equation of motion can be obtained:

$$\frac{\partial \mathcal{L}}{\partial x} = \frac{d}{dt}\frac{\partial \mathcal{L}}{\partial \dot{x}}$$

$$(m_1 \sin\theta - m_2)g = (m_1 + m_2)\ddot{x}$$

$$\ddot{x} = \frac{m_1 \sin\theta - m_2}{m_1 + m_2}g$$

PROBLEM 6.5
A mass m is attached to a spring of constant k, horizontally on a frictionless plane, as in Figure 6.7.

a. Find the Lagrangian and solve for $\ddot{x}$.
b. Using Newton's Second Law, solve for $\ddot{x}$.

SOLUTION 6.5
a. The kinetic energy and potential energy of the system are
$$T = \frac{1}{2} m\dot{x}^2 \text{ and } U = \frac{1}{2} kx^2$$
Subtracting potential energy from kinetic energy outputs the Lagrangian:

$$\mathcal{L} = T - U = \frac{1}{2} m\dot{x}^2 - \frac{1}{2} kx^2$$

Using the relation $\dfrac{\partial \mathcal{L}}{\partial x} = \dfrac{d}{dt} \dfrac{\partial \mathcal{L}}{\partial \dot{x}}$, an expression for $\ddot{x}$ is obtained.

$$-kx = \frac{d}{dt}(m\dot{x})$$

$$-kx = m\ddot{x}$$

$$\ddot{x} = -\frac{k}{m} x$$

b. According to Newton's Second Law, $F_x = m\ddot{x}$. On the block, only the force from the spring is acting in the x direction. Thus, $F_x = -kx$
So:

$$-kx = m\ddot{x}$$

$$\ddot{x} = -\frac{k}{m} x$$

The result is the same as the one obtained in part (a).

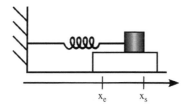

FIGURE 6.7   Mass attached to a spring extended to point x_s. x_e is the equilibrium point of the spring.

PROBLEM 6.6

A particle with initial mass m_0 and initial velocity v_0 begins losing mass according to the equation $m(t) = m_0 e^{-\alpha t}$ where α is constant. If there are no external forces, find an expression for the velocity.

Note: This is the same problem text as Problem 1.7, but now should be solved using Lagrange's equations.

SOLUTION 6.6

Since there is no potential energy in this problem

$$\mathcal{L} = T = \frac{1}{2} m(t)\dot{x}^2 = \frac{1}{2} m_0 e^{-\alpha t}\dot{x}^2$$

Using Lagrange's equations yields

$$\frac{d}{dt}\frac{\partial \mathcal{L}}{\partial \dot{x}} = \frac{\partial \mathcal{L}}{\partial x}$$

$$m_0 \frac{d}{dt}(e^{-\alpha t}\dot{x}) = 0$$

$$m_0(\ddot{x}e^{-\alpha t} - \alpha e^{-\alpha t}\dot{x}) = 0$$

$$m_0 e^{-\alpha t}(\ddot{x} - \alpha \dot{x}) = 0$$

$$\ddot{x} - \alpha \dot{x} = 0$$

This is a separable differential equation.

$$\frac{d\dot{x}}{dt} = \alpha \dot{x}$$

$$\frac{d\dot{x}}{\dot{x}} = \alpha\, dt$$

$$\ln \dot{x} = \alpha t + C$$

$$\dot{x} = A e^{\alpha t}$$

where C and A are constants. Since the initial velocity is v_0,

$$\dot{x}(0) = v_0 = A$$

Therefore, the velocity is given by

$$\dot{x}(t) = v_0 e^{\alpha t}$$

which is exactly what was found in Chapter 1.

PROBLEM 6.7

A small bead of mass m can slide without friction on a parabolic wire of equation $z = ax^2$ rotating with angular velocity ω, as in Figure 6.8.

    a. Find the velocity of the bead.
    b. Find the Lagrangian.
    c. Use Lagrange equation to find the equation of motion.
    d. Discuss the equilibrium position.

SOLUTION 6.7

    a. The cylindrical polar coordinates are the most appropriate in this case, the parabola rotating around the z-axis with the angular velocity $\omega = \dot{\phi}$. Consider the equation of the parabola as $z = ax^2$. The position of the bead is $(x, 0, ax^2)$ and the velocity is calculated following the steps:

$$z = ax^2$$

$$\dot{z} = 2ax\dot{x}$$

$$v^2 = \dot{x}^2 + (x\omega)^2 + \dot{z}^2 = \dot{x}^2 + x^2\omega^2 + (2ax\dot{x})^2 = \dot{x}^2 + x^2\omega^2 + 4a^2x^2\dot{x}^2$$

    b. The Lagrangian is

$$\mathcal{L} = T - U = \frac{m}{2}(\dot{x}^2 + x^2\omega^2 + 4a^2x^2\dot{x}^2) - mgax^2$$

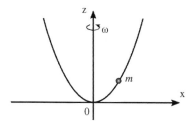

FIGURE 6.8   A bead is sliding without friction on a parabolic wire rotating about the vertical axis with angular velocity ω.

c. Lagrange equation for x

$$\frac{\partial \mathcal{L}}{\partial x} = \frac{d}{dt}\frac{\partial \mathcal{L}}{\partial \dot{x}}$$

$$\frac{\partial \mathcal{L}}{\partial x} = mx\omega^2 + 4ma^2 x\dot{x}^2 - 2mgax$$

$$\frac{\partial \mathcal{L}}{\partial \dot{x}} = m\dot{x} + 4ma^2 x^2 \dot{x}$$

$$\frac{d}{dt}\frac{\partial \mathcal{L}}{\partial \dot{x}} = m\ddot{x} + 4ma^2(2x\dot{x}\dot{x} + x^2\ddot{x}) = m\ddot{x} + 4ma^2(2x\dot{x}^2 + x^2\ddot{x})$$

And by considering both terms of the Lagrangian equation, it yields

$$mx\omega^2 + 4ma^2 x\dot{x}^2 - 2mgax = m\ddot{x} + 4ma^2(2x\dot{x}^2 + x^2\ddot{x})$$

And with a bit of rearrangement and after dividing by mass,

$$\ddot{x} + 4a^2 x^2 \ddot{x} + 4a^2 x\dot{x}^2 = x\omega^2 - 2gax$$

From here, the equation of motion is

$$\ddot{x}(1 + 4a^2 x^2) + 4a^2 x\dot{x}^2 = x(\omega^2 - 2ga)$$

d. If the bead positions itself at height $H = ax^2$ and x is an equilibrium position, it means that $\dot{x} = 0$ and $\ddot{x} = 0$.
   The equation of motion yields

$$x(\omega^2 - 2ga) = 0$$

This can happen if $x = 0$, at the apex of the parabola, or when $\omega^2 - 2ga = 0$, which means that the angular velocity is $\omega = \sqrt{2ga}$.
   In this particular case, the bead can position itself at any height z, and these constitute unstable positions of equilibrium.
   Looking at the case when the bead is at $x = 0$ and checking the stability of the bead, if we consider a position ε very small, close to zero, and $\dot{\varepsilon}$ very small as well, with $\dot{\varepsilon}^2 \to 0$, then the equation of motion becomes

$$\ddot{x} = x(\omega^2 - 2ga)$$

For angular velocities ω smaller than $\sqrt{2ga}$, the bead may go up a bit and then slide back toward the equilibrium position $x = 0$.

For high angular velocities, that is, $\omega > \sqrt{2ga}$, the bead may go further and further from the equilibrium position.

PROBLEM 6.8

A mass m is moving on a frictionless sphere of radius R. The force on the sphere is $\vec{F} = -mg\,\hat{r}$. Find the Lagrangian and solve in polar coordinates.

SOLUTION 6.8

The kinetic energy of the mass is $T = \dfrac{1}{2}mv^2$, which converted in polar coordinates is written as:

$$T = \frac{1}{2}mv^2 = \frac{1}{2}m((r\dot{\phi}\sin\theta)^2 + (r\dot{\theta})^2)$$

The potential energy of the mass is $U = mgr$. The Lagrangian, $\mathcal{L} = T - U$, is obtained, after substituting r by R (the surface of the sphere).

$$\mathcal{L} = \frac{1}{2}m(R^2\dot{\phi}^2\sin^2\theta + R^2\dot{\theta}^2) - mgR$$

Solving for $\ddot{\phi}$ is first done by using the relation

$$\frac{\partial\mathcal{L}}{\partial\phi} = \frac{d}{dt}\frac{\partial\mathcal{L}}{\partial\dot{\phi}}$$

so

$$0 = \frac{d}{dt}(mR^2\dot{\phi}\sin^2\theta)$$

$$0 = mR^2\sin^2\theta\,\ddot{\phi}$$

and $\ddot{\phi} = 0$. Now, an expression for $\ddot{\theta}$ is obtained the same way.

$$\frac{\partial\mathcal{L}}{\partial\theta} = \frac{d}{dt}\frac{\partial\mathcal{L}}{\partial\dot{\theta}}$$

$$mR^2\dot{\phi}^2\cos\theta\sin\theta = mR^2\ddot{\theta}$$

$$\ddot{\theta} = \dot{\phi}^2\cos\theta\sin\theta$$

PROBLEM 6.9

A mass m is suspended vertically on a spring of constant k, as in Figure 6.9. Find the Lagrangian and solve for $\ddot{x}$.

SOLUTION 6.9

The sum of the forces acting on the mass is

$$F = mg - kx$$

The potential energy is $U = -\int F\,dx$. So:

$$U = -mgx + \frac{1}{2}kx^2$$

The kinetic energy of the mass is

$$T = \frac{1}{2}m\dot{x}^2$$

Thus, the Lagrangian is

$$\mathcal{L} = T - U = \frac{1}{2}m\dot{x}^2 + mgx - \frac{1}{2}kx^2$$

Using the Lagrangian, solving for $\ddot{x}$ is straightforward since $\dfrac{\partial \mathcal{L}}{\partial x} = \dfrac{d}{dt}\dfrac{\partial \mathcal{L}}{\partial \dot{x}}$.

$$mg - kx = \frac{d}{dt}(m\dot{x})$$

$$\ddot{x} = g - \frac{k}{m}x$$

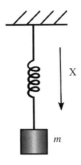

FIGURE 6.9   Mass hanging vertically, attached to a spring.

PROBLEM 6.10

Consider a bead of mass m on the spinning wire as in Figure 6.10. If the wire is in the shape of an upside-down Gaussian, specifically $f(\rho) = -\dfrac{1}{\sqrt{2\pi}} e^{-\frac{\rho^2}{2}}$, find any equilibrium positions and comment on their stability.

SOLUTION 6.10

This type of analysis can be performed using Lagrange's equations which means the kinetic energy T and potential energy U must be found. This requires the position of the bead, which is given by

$$\vec{r} = \rho\,\hat{\rho} + f(\rho)\hat{z} = \rho\,\hat{\rho} - \frac{1}{\sqrt{2\pi}} e^{-\frac{\rho^2}{2}}\,\hat{z}$$

with the velocity given by

$$\vec{v} = \dot{\rho}\,\hat{\rho} + \frac{1}{\sqrt{2\pi}}\rho\dot{\rho}\, e^{-\frac{\rho^2}{2}}\,\hat{z} + \rho\omega\,\hat{\phi}$$

Thus, the kinetic energy is

$$T = \frac{1}{2}mv^2 = \frac{1}{2}m\left(\dot{\rho}^2 + \frac{1}{2\pi}\rho^2\dot{\rho}^2 e^{-\rho^2} + \rho^2\omega^2\right)$$

The potential energy is given by

$$U = mg\left(|f(0)| - |f(\rho)|\right) = mg\left(\frac{1}{\sqrt{2\pi}} - \frac{1}{\sqrt{2\pi}} e^{-\frac{\rho^2}{2}}\right) = \frac{mg}{\sqrt{2\pi}}\left(1 - e^{-\frac{\rho^2}{2}}\right)$$

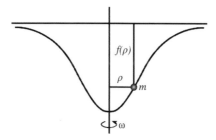

FIGURE 6.10 Bead on a spinning wire.

The Lagrangian is therefore

$$\mathcal{L} = T - U = \frac{1}{2}m\left(\dot{\rho}^2 + \frac{1}{2\pi}\rho^2\dot{\rho}^2 e^{-\rho^2} + \rho^2\omega^2\right) - \frac{mg}{\sqrt{2\pi}}\left(1 - e^{-\frac{\rho^2}{2}}\right)$$

which will be used in Lagrange's equations

$$\frac{d}{dt}\frac{\partial \mathcal{L}}{\partial \dot{\rho}} = \frac{\partial \mathcal{L}}{\partial \rho}$$

Now, to calculate each element

$$\frac{\partial \mathcal{L}}{\partial \dot{\rho}} = \frac{1}{2}m\left(2\dot{\rho} + \frac{1}{\pi}\rho^2\dot{\rho}e^{-\rho^2}\right) = \dot{\rho}m\left(1 + \frac{1}{2\pi}\rho^2 e^{-\rho^2}\right)$$

$$\frac{d}{dt}\frac{\partial \mathcal{L}}{\partial \dot{\rho}} = \ddot{\rho}m\left(1 + \frac{1}{2\pi}\rho^2 e^{-\rho^2}\right) + \dot{\rho}m\frac{d}{dt}\left(1 + \frac{1}{2\pi}\rho^2 e^{-\rho^2}\right)$$

$$= \ddot{\rho}m\left(1 + \frac{1}{2\pi}\rho^2 e^{-\rho^2}\right) + \dot{\rho}m\left(\frac{1}{2\pi}(2\rho\dot{\rho}e^{-\rho^2} + \rho^2 e^{-\rho^2}(-2\rho\dot{\rho}))\right)$$

$$= \ddot{\rho}m\left(1 + \frac{1}{2\pi}\rho^2 e^{-\rho^2}\right) + \frac{\rho\dot{\rho}^2 m}{\pi}e^{-\rho^2}(1 - \rho^2)$$

$$\frac{\partial \mathcal{L}}{\partial \rho} = \frac{1}{2}m\left(\frac{1}{2\pi}\dot{\rho}^2(2\rho\dot{\rho}e^{-\rho^2} + \rho^2 e^{-\rho^2}(-2\rho\dot{\rho})) + 2\rho\omega^2\right) - \frac{mg\rho}{\sqrt{2\pi}}e^{-\frac{\rho^2}{2}}$$

$$= \frac{1}{2}m\left(\frac{1}{\pi}\dot{\rho}^2(\rho\dot{\rho}e^{-\rho^2} - \dot{\rho}\rho^3 e^{-\rho^2}) + 2\rho\omega^2\right) - \frac{mg\rho}{\sqrt{2\pi}}e^{-\frac{\rho^2}{2}}$$

Combining everything yields

$$\ddot{\rho}m\left(1 + \frac{1}{2\pi}\rho^2 e^{-\rho^2}\right) + \frac{\rho\dot{\rho}^2 m}{\pi}e^{-\rho^2}(1 - \rho^2)$$

$$= \frac{1}{2}m\left(\frac{1}{\pi}\dot{\rho}^2(\rho\dot{\rho}e^{-\rho^2} - \dot{\rho}\rho^3 e^{-\rho^2}) + 2\rho\omega^2\right) - \frac{mg\rho}{\sqrt{2\pi}}e^{-\frac{\rho^2}{2}}$$

Since the goal is to analyze the stability, the bead cannot be moving in the ρ direction; therefore,

$$\ddot{\rho} = 0,\ \dot{\rho} = 0$$

and the equation reduces to

$$m\rho\omega^2 - \frac{mg\rho}{\sqrt{2\pi}} e^{-\frac{\rho^2}{2}} = 0$$

$$\rho\left(m\omega^2 - \frac{mg}{\sqrt{2\pi}} e^{-\frac{\rho^2}{2}} \right) = 0$$

From this, it is clear $\rho = 0$ is an equilibrium position. Intuitively, this makes sense; when the mass is at $\rho = 0$, it doesn't experience any force due to the rotation. Considering equilibrium when $\rho \neq 0$ yields

$$m\omega^2 - \frac{mg}{\sqrt{2\pi}} e^{-\frac{\rho^2}{2}} = 0$$

Solving for ω

$$\omega = \sqrt{\frac{g}{\sqrt{2\pi}}} e^{-\frac{\rho^2}{4}}$$

Due to the shape of the wire, the angular velocity must depend on the position. For ρ close to zero,

$$\omega \approx \sqrt{\frac{g}{\sqrt{2\pi}}}$$

As ρ increases, the slope of wire decreases (exponentially); so, in order to keep the bead from moving, the velocity must decrease (exponentially). Very far from $\rho = 0$, where the wire is very flat, an extremely small ω is required.

PROBLEM 6.11
Consider a mass m attached to a pendulum of length l and a spring of constant k as depicted in Figure 6.11. When the mass is at the lowest point of the pendulum, the spring is at its rest length. Considering a small displacement, find the angular velocity.

Note: This is the same problem text as Problem 5.12, but now should be solved using Lagrange's equations.

SOLUTION 6.11
While it may be clear at this point that the velocity of a mass on a pendulum is $l\dot{\phi}$ but to get more practice in deriving it (for cases which aren't so easy), consider the position

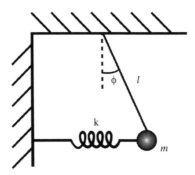

FIGURE 6.11 Mass connected to a pendulum and a spring.

$$\vec{r} = l\sin\phi\,\hat{x} + l\cos\phi\,\hat{y} = l(\sin\phi\,\hat{x} + \cos\phi\,\hat{y})$$

with velocity

$$\vec{v} = l(\cos\phi\,\dot{\phi}\,\hat{x} - \sin\phi\,\dot{\phi}\,\hat{y})$$

so

$$T = \frac{1}{2}mv^2 = \frac{1}{2}m\vec{v}\cdot\vec{v} = \frac{1}{2}ml^2\dot{\phi}^2$$

The potential energy is due to a combination of gravity and the spring

$$U = mgl(1 - \cos\phi) + \frac{1}{2}k(l\tan\phi)^2$$

Considering small angles, this becomes

$$U = \frac{mgl}{2}\phi^2 + \frac{1}{2}kl^2\phi^2$$

Therefore, the Lagrangian is given by

$$\mathcal{L} = T - U = \frac{1}{2}ml^2\dot{\phi}^2 - \frac{mgl}{2}\phi^2 - \frac{1}{2}kl^2\phi^2 = \frac{1}{2}ml^2\dot{\phi}^2 - \frac{l}{2}(mg - kl)\phi^2$$

Using Lagrange's equations yields

$$\frac{d}{dt}\frac{\partial\mathcal{L}}{\partial\dot{\phi}} = \frac{\partial\mathcal{L}}{\partial\phi}$$

$$ml^2\ddot{\phi} = -l(mg - kl)\phi$$

$$\ddot{\phi} = -\frac{mg - kl}{ml}$$

Therefore, the angular velocity is

$$\omega = \sqrt{\frac{mg - kl}{ml}}$$

PROBLEM 6.12

Consider a block of mass m on ramp at an incline θ and connected to a spring of constant k, as shown in Figure 6.12. Using Lagrange's equations, find the angular frequency of oscillations following a displacement of the block.

SOLUTION 6.12

Considering the x-axis is parallel to the ramp, the kinetic energy is given by

$$T = \frac{1}{2}m\dot{x}^2$$

Without worrying about the true height of the ramp, the potential energy is given by

$$U = \frac{1}{2}k(x + d)^2 + mg(\text{const} - x\sin\theta)$$

where d is the "pre-stretched" displacement of the spring due to the mass but not due to the oscillations. This quantity is found by considering Figure 6.13
Therefore,

$$kd = mg\sin\theta$$

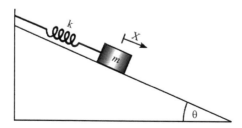

FIGURE 6.12 Block on a ramp connected to a spring.

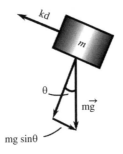

FIGURE 6.13   Free body diagram of the block.

The Lagrangian is given by

$$\mathcal{L} = T - U = \frac{1}{2}m\dot{x}^2 - \frac{1}{2}k(x+d)^2 - mg(\text{const} - x\sin\theta)$$

Using Lagrange's equations yields

$$\frac{d}{dt}\frac{\partial\mathcal{L}}{\partial\dot{x}} = \frac{\partial\mathcal{L}}{\partial x}$$

$$m\ddot{x} = -(k(x+d) - mg\sin\theta)$$

$$m\ddot{x} = -(kx + kd - mg\sin\theta)$$

Notice $kd - mg\sin\theta = 0$, so this reduces to

$$\ddot{x} = -\frac{k}{m}x$$

and the angular velocity is

$$\omega = \sqrt{\frac{k}{m}}$$

An interesting note: The angular frequency of oscillation does not depend on θ at all.

PROBLEM 6.13

Consider the pendulum–spring system from Figure 6.14 consisting of a mass m connected to a pendulum of length l and a spring with constant k and rest length l. Assuming a small angular displacement, find the angular frequency of the oscillations.

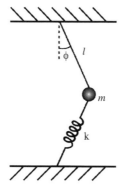

FIGURE 6.14   Mass connected to a pendulum above and a spring below.

SOLUTION 6.13

This can be solved using Lagrange's equations, with kinetic energy given by

$$T = \frac{1}{2}ml^2\dot{\phi}^2$$

In order to find the potential energy associated with the spring, consider Figures 6.15 and 6.16, where d is the amount the spring stretches. Using the law of cosines yields

$$(l+d)^2 = (2l)^2 + l^2 - 2(2l)(l)\cos\phi$$

$$(l+d)^2 = 4l^2 + l^2 - 4l^2 \cos\phi$$

$$(l+d)^2 = l^2(1 + 4(1 - \cos\phi))$$

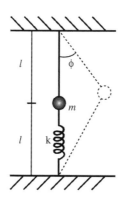

FIGURE 6.15   Projection of the mass, pendulum, and spring upon displacement.

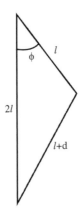

FIGURE 6.16   Triangle illustrating distances involved in the displacement.

$$l + d = l\sqrt{1 + 4(1 - \cos\phi)}$$

$$d = l\left(\sqrt{1 + 4(1 - \cos\phi)} - 1\right)$$

Therefore, the potential energy is given by

$$U = mgl(1 - \cos\phi) + \frac{1}{2}kl^2\left(\sqrt{1 + 4(1 - \cos\phi)} - 1\right)^2$$

and the Lagrangian is

$$\mathcal{L} = T - U = \frac{1}{2}ml^2\dot{\phi}^2 - mgl(1 - \cos\phi) - \frac{1}{2}kl^2(\sqrt{1 + 4(1 - \cos\phi)} - 1)^2$$

Considering a small angle approximation

$$1 - \cos\phi \approx \frac{\phi^2}{2}$$

so

$$\mathcal{L} = \frac{1}{2}ml^2\dot{\phi}^2 - \frac{mgl}{2}\phi^2 - \frac{1}{2}kl^2(\sqrt{1 + 2\phi^2} - 1)^2$$

Also notice

$$\sqrt{1 + 2\phi^2} - 1 \approx 1 + \phi^2 - 1 = \phi^2$$

Therefore,

$$\mathcal{L} = \frac{1}{2}ml^2\dot{\phi}^2 - \frac{mgl}{2}\phi^2 - \frac{1}{2}kl^2\phi^4$$

Since $\phi \ll 1$, $\phi^4 \ll \phi^2$ and, provided the spring constant isn't massive, the last term in the Lagrangian can be omitted; essentially, the spring does not stretch enough in the small angle approximation to affect the angular frequency. The Lagrangian then becomes

$$\mathcal{L} = \frac{1}{2}ml^2\dot{\phi}^2 - \frac{mgl}{2}\phi^2$$

Using Lagrange's equations yields

$$\frac{d}{dt}\frac{\partial\mathcal{L}}{\partial\dot{\phi}} = \frac{\partial\mathcal{L}}{\partial\phi}$$

$$ml^2\ddot{\phi} = -lmg\phi$$

$$\ddot{\phi} = -\frac{g}{l}$$

and the angular frequency of oscillation is

$$\omega = \sqrt{\frac{g}{l}}$$

Note: In the case of a massive spring constants, it is possible ϕ^2 and $k\phi^4$ are of similar order, and the last term in the Lagrangian could no longer be omitted. This would have resulted in the differential equation

$$ml^2\ddot{\phi} = -lmg\phi - 2kl^2\phi^3$$

which is a nonlinear differential equation and could be solved numerically.

7 Hamiltonian Formalism

7.1 THEORY

This chapter presents the Hamiltonian formalism. Many problems solved in Chapter 6 using Lagrangian formalism are solved here by using the Hamiltonian formalism. It is important to note that the Hamiltonian formalism is based on Lagrangian formalism.

7.1.1 THE HAMILTONIAN

The Lagrangian is obtained as

$$\mathcal{L}(q_1,\ldots,q_n,\dot{q}_1,\ldots,\dot{q}_n,t) = T - U$$

The Hamiltonian is defined as

$$\mathcal{H} = \sum_{i=1}^{n} p_i \dot{q}_i - \mathcal{L}$$

where the generalized momenta are

$$p_i = \frac{\partial \mathcal{L}}{\partial \dot{q}_i}$$

The next step is to express the Hamiltonian function as a function of the variables q_i and p_i. For this, the generalized velocities $\dot{q}_i$ are written in terms of the coordinates q_i and p_i and the Hamiltonian is written in terms of q_i and p_i, that is, $\mathcal{H}(q_1,\ldots,q_n,p_1,\ldots,p_n,t)$.

Hamilton's equations are, with i from 1 to n,

$$\dot{q}_i = \frac{\partial \mathcal{H}}{\partial p_i}$$

$$\dot{p}_i = -\frac{\partial \mathcal{H}}{\partial q_i}$$

7.1.2 EXAMPLE – ONE-DIMENSIONAL SYSTEMS

$$\mathcal{L} = \mathcal{L}(q, \dot{q}) = T - U$$

DOI: 10.1201/9781003365709-7

The generalized momentum is

$$p = \frac{\partial \mathcal{L}}{\partial \dot{q}}$$

and it is substituted in the Hamiltonian,

$$\mathcal{H} = p\dot{q} - \mathcal{L}$$

Hamilton's equations for a one-dimensional system are

$$\dot{q} = \frac{\partial \mathcal{H}}{\partial p}$$

and

$$\dot{p} = -\frac{\partial \mathcal{H}}{\partial q}$$

7.2   PROBLEMS AND SOLUTIONS

PROBLEM 7.1
A mass m is moving in xyz coordinates. A force F is acting on the mass. For the following cases, get the Hamiltonian and find Hamilton's equations.

a. $\vec{F} = ax\,\hat{x}$
b. $\vec{F} = ay\,\hat{x} + ax\,\hat{y}$
c. $\vec{F} = a\,\hat{x} - by\,\hat{y} + cz\,\hat{z}$

SOLUTION 7.1
a. First, the Lagrangian is determined:

$$U = -\int F\,dx = -\frac{1}{2}ax^2$$

$$\mathcal{L} = T - U = \frac{1}{2}m(\dot{x}^2 + \dot{y}^2 + \dot{z}^2) + \frac{1}{2}ax^2$$

The momentum in each direction is obtained:

$$p_x = \frac{\partial L}{\partial \dot{x}} = m\dot{x}$$

$$p_y = \frac{\partial L}{\partial \dot{y}} = m\dot{y}$$

$$p_z = \frac{\partial L}{\partial \dot{z}} = m\dot{z}$$

Using the expressions of momentum and the Lagrangian, the Hamiltonian is

$$\mathcal{H} = p_x\dot{x} + p_y\dot{y} + p_z\dot{z} - \frac{1}{2}m(\dot{x}^2 + \dot{y}^2 + \dot{z}^2) - \frac{1}{2}ax^2$$

$$= \frac{p_x^2}{m} + \frac{p_y^2}{m} + \frac{p_z^2}{m} - \frac{1}{2}m\left(\frac{p_x^2}{m^2} + \frac{p_y^2}{m^2} + \frac{p_z^2}{m^2}\right) - \frac{1}{2}ax^2$$

$$= \frac{1}{2m}(p_x^2 + p_y^2 + p_z^2) - \frac{1}{2}ax^2$$

Deriving the Hamiltonian leads to Hamilton's equations.

$$\dot{x} = \frac{\partial \mathcal{H}}{\partial p_x} = \frac{p_x}{m} \qquad \dot{p}_x = \frac{-\partial \mathcal{H}}{\partial x} = +ax$$

$$\dot{y} = \frac{\partial \mathcal{H}}{\partial p_y} = \frac{p_y}{m} \qquad \dot{p}_y = \frac{-\partial \mathcal{H}}{\partial y} = 0$$

$$\dot{z} = \frac{\partial \mathcal{H}}{\partial p_z} = \frac{p_z}{m} \qquad \dot{p}_z = \frac{-\partial \mathcal{H}}{\partial z} = 0$$

b. The process is the same as in part (a).
First, the Lagrangian is determined:

$$U = -axy$$

Checking that the potential energy U is correctly calculated can be achieved by taking its gradient:

$$\vec{F} = -\nabla U = -\frac{\partial U}{\partial x}\hat{x} - \frac{\partial U}{\partial y}\hat{y} - \frac{\partial U}{\partial z}\hat{z} = ay\,\hat{x} + ax\,\hat{y}$$

$$\mathcal{L} = \frac{1}{2}m(\dot{x}^2 + \dot{y}^2 + \dot{z}^2) + axy$$

The momentum in each direction is obtained:

$$p_x = m\dot{x}$$

$$p_y = m\dot{y}$$

$$p_z = m\dot{z}$$

Using the expressions of momentum and the Lagrangian, the Hamiltonian is

$$\mathcal{H} = \frac{1}{2m}(p_x^2 + p_y^2 + p_z^2) - axy$$

Deriving the Hamiltonian leads to Hamilton's equations:

$$\dot{x} = \frac{p_x}{m} \qquad \dot{p}_x = ay$$

$$\dot{y} = \frac{p_y}{m} \qquad \dot{p}_y = ax$$

$$\dot{z} = \frac{p_z}{m} \qquad \dot{p}_z = 0$$

c. The process is the same as in parts (a) and (b).
   First, the Lagrangian is determined:

$$U = -ax + \frac{1}{2}by^2 - \frac{1}{2}cz^2$$

$$\mathcal{L} = \frac{1}{2}m(\dot{x}^2 + \dot{y}^2 + \dot{z}^2) + ax - \frac{1}{2}by^2 + \frac{1}{2}cz^2$$

The momentum in each direction is obtained:

$$p_x = m\dot{x}$$

$$p_y = m\dot{y}$$

$$p_z = m\dot{z}$$

Using the expressions of momentum and the Lagrangian, the Hamiltonian is

$$\mathcal{H} = \frac{1}{2m}(p_x^2 + p_y^2 + p_z^2) - ax + \frac{1}{2}by^2 - \frac{1}{2}cz^2$$

Deriving the Hamiltonian leads to Hamilton's equations:

$$\dot{x} = \frac{p_x}{m} \qquad \dot{p}_x = a$$

$$\dot{y} = \frac{p_y}{m} \qquad \dot{p}_y = -by$$

$$\dot{z} = \frac{p_z}{m} \qquad \dot{p}_x = cz$$

PROBLEM 7.2

A mass is attached to a spring of constant k, horizontally on a frictionless plane

a. Find the Hamiltonian and Hamilton's equations.
b. Find the equation of motion.

SOLUTION 7.2

a. The Lagrangian

$$\mathcal{L} = T - U = \frac{1}{2}m\dot{x}^2 - \frac{1}{2}kx^2$$

is used to find the Hamiltonian:

$$p_x = m\dot{x}$$

$$\mathcal{H} = \frac{p_x^2}{2m} + \frac{1}{2}kx^2$$

Now, Hamilton's equations can be solved by derivation:

$$\dot{p}_x = \frac{-\partial \mathcal{H}}{\partial x} = -kx$$

$$\dot{x} = \frac{\partial \mathcal{H}}{\partial p_x} = \frac{p_x}{m}$$

b. From Hamilton's equations obtained in part (a), the momentum of the mass is $p_x = m\dot{x}$.
After derivation,

$$\dot{p}_x = m\ddot{x}$$

Again, from results obtained in part (a), $\dot{p}_x = -kx$. So

$$m\ddot{x} = -kx$$

PROBLEM 7.3

Consider the Atwood machine in Figure 7.1: two masses m_1 and m_2 suspended by an inextensible string over a frictionless pulley.

    a. Obtain the Lagrangian.

    b. Find the generalized momentum by using $p = \dfrac{\partial \mathcal{L}}{\partial \dot{x}}$ and obtain the Hamiltonian

      $\mathcal{H} = p\dot{q} - \mathcal{L}$ for one degree of freedom.

    c. Use Hamilton's equations $\dot{x} = \dfrac{\partial \mathcal{H}}{\partial p}$ and $\dot{p} = -\dfrac{\partial \mathcal{H}}{\partial x}$ to obtain the acceleration of the system.

Note: This is the same problem as 6.1, but now it is solved using Hamilton's equations.

SOLUTION 7.3

    a. As in the similar problem from Chapter 6 (Problem 6.1), the length of the string is constant, so $x + y + \pi R = constant$, so $y = -x + \pi R$, and $\dot{y} = -\dot{x}$ and $\dot{y}^2 = \dot{x}^2$.
      The kinetic energy is

$$T = \frac{m_1 + m_2}{2} \dot{x}^2$$

The potential energy is

$$U = -m_1 gx - m_2 gy = -(m_1 - m_2)gx + constant$$

The Lagrangian is

$$\mathcal{L} = T - U = \frac{m_1 + m_2}{2} \dot{x}^2 + (m_1 - m_2)gx$$

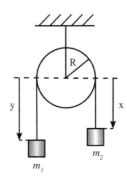

FIGURE 7.1 Atwood machine with two masses.

b. The generalized momentum is

$$p = \frac{\partial \mathcal{L}}{\partial \dot{x}} = (m_1 + m_2)\dot{x}$$

The next step is to write $\dot{x}$ in terms of the generalized momentum:

$$\dot{x} = \frac{p}{(m_1 + m_2)}$$

The Lagrangian in terms of x and p becomes

$$\mathcal{L}(x, p) = \frac{(m_1 + m_2)}{2} \left(\frac{p}{m_1 + m_2} \right)^2 + (m_1 - m_2)gx$$

$$= \frac{p^2}{2(m_1 + m_2)} + (m_1 - m_2)gx$$

The Hamiltonian is obtained by using the expression $\mathcal{H} = p\dot{q} - \mathcal{L}$.

$$\mathcal{H} = p\frac{p}{m_1 + m_2} - \frac{p^2}{2(m_1 + m_2)} - (m_1 - m_2)gx$$

$$= \frac{p^2}{2(m_1 + m_2)} - (m_1 - m_2)gx$$

c. Hamilton's equations are

$$\dot{x} = \frac{\partial \mathcal{H}}{\partial p} = \frac{p}{m_1 + m_2} \tag{7.1}$$

$$\dot{p} = -\frac{\partial \mathcal{H}}{\partial x} = (m_1 - m_2)g \tag{7.2}$$

By taking the derivative of $\dot{x}$ with respect to time in the first Hamilton Equation (7.1) and by substituting $\dot{p}$ from the second Hamilton Equation (7.2), the acceleration is obtained

$$\ddot{x} = \frac{\dot{p}}{m_1 + m_2} = \frac{(m_1 - m_2)}{m_1 + m_2}g$$

Which is the same result from Chapter 6, as expected.

PROBLEM 7.4

Consider a block of mass m sliding down a frictionless ramp at an incline θ as in Figure 7.2. Find the velocity of the block at time t if the block is stationary at $t = 0$ and x is measured parallel to the incline.

Note: This is the same problem text as Problem 6.3 but now should be solved using Hamilton's equations.

SOLUTION 7.4

In order to use Hamilton's equations, the Lagrangian is required. The kinetic energy is simply

$$T = \frac{1}{2}m\dot{x}^2$$

with the potential given by:

$$U = mg(\text{const} - x\sin\theta)$$

Therefore, the Lagrangian is

$$\mathcal{L} = T - U = \frac{1}{2}m\dot{x}^2 - mg(\text{const} - x\sin\theta)$$

To use this in Hamilton's equations, the velocity term must be substituted for the generalized momentum. So

$$p = \frac{\partial \mathcal{L}}{\partial \dot{x}} = m\dot{x}$$

$$\dot{x} = \frac{p}{m}$$

The Lagrangian can now be rewritten in terms of x and p, yielding

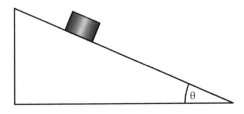

FIGURE 7.2   Block placed on a ramp.

$$\mathcal{L} = \frac{p^2}{2m} - mg(\text{const} - x\sin\theta)$$

The Hamiltonian is then given by

$$\mathcal{H} = p\dot{x} - \mathcal{L} = \frac{p^2}{m} - \frac{p^2}{2m} + mg(\text{const} - x\sin\theta) = \frac{p^2}{2m} + mg(\text{const} - x\sin\theta)$$

Hamilton's equations yield

$$\dot{x} = \frac{\partial\mathcal{H}}{\partial p} = \frac{p}{m}$$

and

$$\dot{p} = -\frac{\partial\mathcal{H}}{\partial x} = mg\sin\theta$$

Therefore,

$$p = \int \dot{p}\,dt = mgt\sin\theta + C$$

where C is a constant. This can be plugged into the equation for $\dot{x}$

$$\dot{x} = \frac{p}{m} = gt\sin\theta + C$$

Since $\dot{x}(0) = 0$, $C = 0$ and

$$\dot{x}(t) = gt\sin\theta$$

as was found before.

PROBLEM 7.5
A mass is moving on a frictionless sphere of radius R. The force on the sphere is $\vec{F} = -mg\hat{r}$.

Find the Hamiltonian and Hamilton's equations.

SOLUTION 7.5
As shown in Problem 6.8, the Lagrangian is

$$\mathcal{L} = \frac{1}{2}m(R^2\dot{\phi}^2\sin^2\theta + R^2\dot{\theta}^2) - mgR$$

The momentum and first derivative for each variable are formulated:

$$p_\phi = \frac{\partial L}{\partial \dot{\phi}} = mR^2\dot{\phi}\sin^2\theta$$

$$\dot{\phi} = \frac{p_\phi}{mR^2\sin^2\theta}$$

and

$$p_\theta = \frac{\partial L}{\partial \dot{\theta}} = mR^2\dot{\theta}$$

$$\dot{\theta} = \frac{p_\theta}{mR^2}$$

Now, the Hamiltonian can be found

$$\mathcal{H} = \frac{p_\phi^2}{mR^2\sin^2\theta} + \frac{p_\theta^2}{mR^2} - \frac{1}{2m}\left(R^2\frac{p_\phi^2}{m^2R^4\sin^4\theta}\sin^2\theta + R^2\frac{p_\theta^2}{m^2R^4}\right) + mgR$$

$$\mathcal{H} = \frac{p_\phi^2}{2mR^2\sin^2\theta} + \frac{p_\theta^2}{2mR^2} + mgR$$

Deriving the Hamiltonian gives Hamilton's equations for ϕ and θ.

$$\dot{\phi} = \frac{\partial \mathcal{H}}{\partial p_\phi} = \frac{p_\phi}{mR^2\sin^2\theta}$$

$$\dot{p}_\phi = -\frac{\partial \mathcal{H}}{\partial \phi} = 0$$

and

$$\dot{\theta} = \frac{\partial \mathcal{H}}{\partial p_\theta} = \frac{p_\theta}{mR^2}$$

$$\dot{p}_\theta = -\frac{\partial \mathcal{H}}{\partial \theta} = \frac{p_\phi^2\cos\theta}{mR^2\sin^3\theta}$$

PROBLEM 7.6

Consider the following system of masses m_1 and m_2, connected by an inextensible massless string as in Figure 7.3. The mass m_1 is on a table and connected to the wall by a spring of spring constant k.

a. Obtain the Lagrangian of the system.
b. Use Lagrange equation to obtain the equation of motion.
c. Find the Hamiltonian, write the Hamilton's equation, and compare the equation obtained for acceleration.

SOLUTION 7.6

a. As discussed in previous problems involving the Atwood machine, since the masses, m_1 and m_2, are connected with an inextensible string, the kinetic energy is

$$T = \frac{m_1 + m_2}{2} \dot{x}^2$$

and the potential energy is

$$U = -m_2 g x + \frac{kx^2}{2}$$

so the Lagrangian is

$$\mathcal{L} = T - U = \frac{m_1 + m_2}{2} \dot{x}^2 + m_2 g x - \frac{kx^2}{2}$$

b. Lagrange equation for x is

$$\frac{\partial \mathcal{L}}{dx} = \frac{d}{dt} \frac{\partial \mathcal{L}}{d\dot{x}}$$

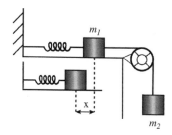

FIGURE 7.3 System of two masses connected by an inextensible string. The system is connected to the wall by a spring of spring constant k. The spring below is simply showing the equilibrium position.

$$\frac{\partial \mathcal{L}}{dx} = m_2 g - kx$$

$$\frac{\partial \mathcal{L}}{d\dot{x}} = (m_1 + m_2)\dot{x}$$

$$\frac{d}{dt}\frac{\partial \mathcal{L}}{d\dot{x}} = (m_1 + m_2)\ddot{x}$$

so

$$(m_1 + m_2)\ddot{x} = m_2 g - kx$$

$$(m_1 + m_2)\ddot{x} + kx - m_2 g = 0$$

c. The generalized momentum is

$$p = \frac{\partial \mathcal{L}}{\partial \dot{x}} = (m_1 + m_2)\dot{x}$$

so

$$\dot{x} = \frac{p}{(m_1 + m_2)}$$

By substituting this into the Lagrangian,

$$\mathcal{L}(x, p) = \frac{p^2}{2(m_1 + m_2)} + m_2 gx - \frac{kx^2}{2}$$

The Hamiltonian is in this case defined as

$$\mathcal{H} = p\dot{q} - \mathcal{L} = p\frac{p}{m_1 + m_2} - \frac{p^2}{2(m_1 + m_2)} - m_2 gx + \frac{kx^2}{2} = \frac{p^2}{2(m_1 + m_2)} - m_2 gx + \frac{kx^2}{2}$$

Hamilton's equations are

$$\dot{x} = \frac{\partial \mathcal{H}}{\partial p} = \frac{p}{m_1 + m_2}$$

$$\dot{p} = -\frac{\partial \mathcal{H}}{\partial x} = m_2 g - kx$$

Deriving again the velocity with respect to time, the acceleration is

$$\ddot{x} = \frac{\dot{p}}{m_1 + m_2} = \frac{m_2 g - kx}{m_1 + m_2}$$

Which is the same expression as before.

$$(m_1 + m_2)\ddot{x} + kx - m_2 g = 0$$

PROBLEM 7.7

Consider a block of mass m on ramp at an incline θ and connected to a spring of constant k, as shown in Figure 7.4. Using Hamilton's equations, find the angular frequency of oscillations following a displacement of the block.

Note: This is the same problem text as Problem 6.12 but now should be solved using Hamilton's equations.

SOLUTION 7.7

In order to use Hamilton's equations, the Lagrangian is required. The kinetic energy is given by

$$T = \frac{1}{2}m\dot{x}^2$$

and the potential energy (without worrying about the true height of the ramp) is given by

$$U = \frac{1}{2}k(x+d)^2 + mg(\text{const} - x\sin\theta)$$

where d is the "pre-stretched" displacement of the spring due to the mass but not due to the oscillations. This was found to be

$$d = \frac{mg\sin\theta}{k}$$

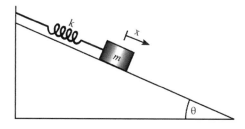

FIGURE 7.4   Block on a ramp connected to a spring.

in Chapter 6. Therefore, the Lagrangian is given by

$$\mathcal{L} = T - U = \frac{1}{2}m\dot{x}^2 - \frac{1}{2}k(x+d)^2 - mg(\text{const} - x\sin\theta)$$

To use this in Hamilton's equations, the velocity term must be substituted for the generalized momentum. So

$$p = \frac{\partial \mathcal{L}}{\partial \dot{x}} = m\dot{x}$$

$$\dot{x} = \frac{p}{m}$$

Now, rewriting the Lagrangian in terms of x and p yields

$$\mathcal{L} = \frac{p^2}{2m} - \frac{1}{2}k(x+d)^2 - mg(\text{const} - x\sin\theta)$$

The Hamiltonian is thus given by

$$\mathcal{H} = p\dot{x} - \mathcal{L} = \frac{p^2}{m} - \frac{p^2}{2m} + \frac{1}{2}k(x+d)^2 + mg(\text{const} - x\sin\theta)$$

$$\mathcal{H} = \frac{p^2}{2m} + \frac{1}{2}k(x+d)^2 + mg(\text{const} - x\sin\theta)$$

Considering Hamilton's equations, the angular velocity can be found via

$$\dot{x} = \frac{\partial \mathcal{H}}{\partial p} = \frac{p}{m}$$

$$\ddot{x} = \frac{\dot{p}}{m}$$

where $\dot{p}$ comes from the other of Hamilton's equations

$$\dot{p} = -\frac{\partial \mathcal{H}}{\partial x} = -k(x+d) + mg\sin\theta = -k\left(x + \frac{mg\sin\theta}{k}\right) + mg\sin\theta = -kx$$

Therefore,

$$\ddot{x} = \frac{\dot{p}}{m} = -\frac{k}{m}x$$

and the angular velocity is given by

$$\omega = \sqrt{\frac{k}{m}}$$

as was found in Chapter 6.

PROBLEM 7.8

Consider an object of mass m connected to the ceiling by a spring, as in Figure 7.5. The system sphere-spring can only move in a vertical plane.

    a. Choose the appropriate coordinates and write the kinetic and the potential energies.
    b. Find the Lagrangian.
    c. Obtain the equation of motion from Lagrange equations.
    d. Write the Hamiltonian and Hamilton's equations and compare with part (c).

SOLUTION 7.8

    a. The spring is moving in a vertical plane and forming the angle θ with the vertical as in Figure 7.6. The mass m is positioned at the length $l + x$ from the fixed point on the ceiling, where x is the distance from the equilibrium position.

    The position of the mass m is $((l + x)\sin\theta, (l + x)\cos\theta)$. The velocity is $v^2 = \dot{x}^2 + [(l + x)\dot{\theta}]^2$ and the kinetic energy is

$$T = \frac{mv^2}{2} = \frac{m}{2}(\dot{x}^2 + (l + x)^2\dot{\theta}^2)$$

The potential energy is both gravitational and elastic,

$$U = -mg(l + x)\cos\theta + \frac{kx^2}{2}$$

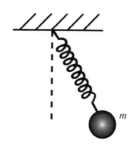

FIGURE 7.5   A sphere of mass m connected to the ceiling by a spring and moving only in a vertical plane.

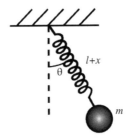

FIGURE 7.6 A sphere of mass m connected to the ceiling by a spring and moving only in a vertical plane, with the angle with respect to the vertical being θ, and the length of the spring length $l + x$.

b. The Lagrangian is as follows:

$$\mathcal{L} = T - U = \frac{m}{2}(\dot{x}^2 + (l+x)^2\dot{\theta}^2) + mg(l+x)\cos\theta - \frac{kx^2}{2}$$

c. The next step is to write the Lagrange equations for x and θ and to calculate them. We start with x:

$$\frac{\partial \mathcal{L}}{\partial x} = \frac{d}{dt}\frac{\partial \mathcal{L}}{\partial \dot{x}}$$

$$\frac{\partial \mathcal{L}}{\partial x} = m(l+x)\dot{\theta}^2 + mg\cos\theta - kx$$

$$\frac{\partial \mathcal{L}}{\partial \dot{x}} = m\dot{x}$$

$$\frac{d}{dt}\frac{\partial \mathcal{L}}{\partial \dot{x}} = m\ddot{x}$$

$$m\ddot{x} = m(l+x)\dot{\theta}^2 + mg\cos\theta - kx \qquad (7.3)$$

$$\ddot{x} = (l+x)\dot{\theta}^2 + g\cos\theta - \frac{k}{m}x$$

Now, the equation for θ:

$$\frac{\partial \mathcal{L}}{\partial \theta} = \frac{d}{dt}\frac{\partial \mathcal{L}}{\partial \dot{\theta}}$$

$$\frac{\partial \mathcal{L}}{\partial \theta} = -mg(l+x)\sin\theta$$

$$\frac{\partial \mathcal{L}}{\partial \dot{\theta}} = m(l+x)^2 \dot{\theta}$$

$$\frac{d}{dt}\frac{\partial \mathcal{L}}{\partial \dot{\theta}} = m(l+x)^2 \ddot{\theta} + 2m(l+x)\dot{\theta}\dot{x}$$

$$-mg(l+x)\sin\theta = m(l+x)^2 \ddot{\theta} + 2m(l+x)\dot{\theta}\dot{x}$$

$$(l+x)\ddot{\theta} + 2\dot{\theta}\dot{x} + g\sin\theta = 0$$

d. The Hamiltonian is obtained by

$$\mathcal{H} = \sum_{i=1}^{n} p_i \dot{q}_i - \mathcal{L} = p_x \dot{q}_x + p_\theta \dot{q}_\theta - \mathcal{L}$$

The generalized momenta are

$$p_x = \frac{\partial \mathcal{L}}{\partial \dot{x}} = m\dot{x}$$

$$\dot{x} = \frac{p_x}{m}$$

and

$$p_\theta = \frac{\partial \mathcal{L}}{\partial \dot{\theta}} = m(l+x)^2 \dot{\theta}$$

$$\dot{\theta} = \frac{p_\theta}{m(l+x)^2}$$

By substituting the generalized momenta, the new form of the Lagrangian is obtained as

$$\mathcal{L} = \frac{m}{2}\left(\frac{p_x^2}{m^2} + (l+x)^2 \frac{p_\theta^2}{m^2(l+x)^4}\right) + mg(l+x)\cos\theta - \frac{kx^2}{2}$$

$$= \frac{p_x^2}{2m} + \frac{p_\theta^2}{2m(l+x)^2} + mg(l+x)\cos\theta - \frac{kx^2}{2}$$

The Hamiltonian is obtained by subsequent substitutions in the formula

$$\mathcal{H} = p_x \dot{q}_x + p_\theta \dot{q}_\theta - \mathcal{L} = \frac{p_x^2}{m} + \frac{p_\theta^2}{m(l+x)^2} - \frac{p_x^2}{2m} - \frac{p_\theta^2}{2m(l+x)^2}$$

$$-mg(l+x)\cos\theta + \frac{kx^2}{2} = \frac{p_x^2}{2m} + \frac{p_\theta^2}{2m(l+x)^2} - mg(l+x)\cos\theta + \frac{kx^2}{2}$$

$$\dot{x} = \frac{\partial \mathcal{H}}{\partial p_x} = \frac{p_x}{m}$$

$$\ddot{x} = \frac{\dot{p}_x}{m}$$

$$\dot{p}_x = -\frac{\partial \mathcal{H}}{\partial x} = \frac{p_\theta^2}{m(l+x)^3} + mg\cos\theta - kx$$

$$\ddot{x} = \frac{p_\theta^2}{m^2(l+x)^3} + g\cos\theta - \frac{k}{m}x$$

Recalling that

$$\dot{\theta} = \frac{p_\theta}{m(l+x)^2}$$

Equation (7.3) from part (c)

$$m\ddot{x} = m(l+x)\dot{\theta}^2 + mg\cos\theta - kx$$

becomes

$$\ddot{x} = \frac{(l+x)p_\theta^2}{m^2(l+x)^4} + g\cos\theta - \frac{k}{m}x$$

which is the same one as using Hamiltonians

$$\ddot{x} = \frac{p_\theta^2}{m^2(l+x)^3} + g\cos\theta - \frac{k}{m}x$$

Now, checking the equation in θ:

$$\dot{\theta} = \frac{\partial \mathcal{H}}{\partial p_\theta} = \frac{p_\theta}{m(l+x)^2}$$

$$\ddot{\theta} = \frac{\dot{p}_\theta}{m(l+x)^2} - \frac{2p_\theta}{m(l+x)^3}\dot{x}$$

$$\dot{p}_\theta = -\frac{\partial \mathcal{H}}{\partial \theta} = -mg(l+x)\sin\theta$$

Therefore,

$$\ddot{\theta} = \frac{-mg(l+x)\sin\theta}{m(l+x)^2} - \frac{2p_\theta}{m(l+x)^3}\dot{x}$$

$$\ddot{\theta} = \frac{-g\sin\theta}{l+x} - \frac{2p_\theta p_x}{m^2(l+x)^3}$$

Recalling that $\dot{\theta} = \frac{p_\theta}{m(l+x)^2}$ and $\dot{x} = \frac{p_x}{m}$ and this is how Equation (7.3) from part (c) is obtained, similarly as by Lagrangian formalism

$$(l+x)\ddot{\theta} + 2\dot{\theta}\dot{x} + g\sin\theta = 0$$

PROBLEM 7.9

A mass m is moving inside and against the side of a half sphere of radius R as in Figure 7.7. The force on the mass is $\vec{F} = -kr\,\hat{r}$, directed toward the bottom of the sphere. Find the Hamiltonian and Hamilton's equations.

SOLUTION 7.9

The kinetic energy of the mass is (Figure 7.8)

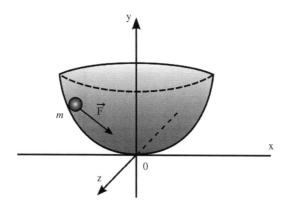

FIGURE 7.7 Mass moving inside a half sphere, with force F acting on it.

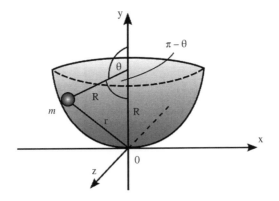

FIGURE 7.8   Mass moving inside a half sphere, with force F acting on it.

$$T = \frac{1}{2}mv^2 = \frac{1}{2}m(R^2\dot{\phi}^2\sin^2\theta + R^2\dot{\theta}^2)$$

The potential energy is

$$U = \frac{1}{2}kr^2$$

Using geometric relations and the properties of isosceles triangles, r is expressed as

$$r = 2R\sin\left(\frac{\pi-\theta}{2}\right)$$

Thus,

$$U = \frac{1}{2}k4R^2\sin^2\left(\frac{\pi-\theta}{2}\right)$$

The Lagrangian of the system is obtained:

$$\mathcal{L} = T - U = \frac{1}{2}m(R^2\dot{\phi}^2\sin^2\theta + R^2\dot{\theta}^2) - 2kR^2\sin^2\left(\frac{\pi-\theta}{2}\right)$$

The momentum corresponding to each variable is obtained:

$$p_\phi = \frac{\partial L}{\partial\dot{\phi}} = mR^2\sin^2\theta\,\dot{\phi}$$

$$p_\theta = \frac{\partial L}{\partial \theta} = mR^2 \dot{\theta}$$

Now, it is possible to find the Hamiltonian $\mathcal{H}$, and Hamilton's equations:

$$\mathcal{H} = \frac{p_\phi^2}{2mR^2 \sin^2 \theta} + \frac{p_\theta^2}{2mR^2} + 2kR^2 \sin^2\left(\frac{\pi - \theta}{2}\right)$$

$$\dot{p}_\phi = -\frac{\partial \mathcal{H}}{\partial \phi} = 0$$

$$\dot{\phi} = \frac{p_\phi}{mR^2 \sin^2 \theta}$$

$$\dot{p}_\theta = -\frac{\partial \mathcal{H}}{\partial \theta} = \frac{\dot{p}_\phi \cos \theta}{mR^2 \sin^2 \theta} + 2kR^2 \cos\left(\frac{\pi - \theta}{2}\right) \sin\left(\frac{\pi - \theta}{2}\right)$$

$$\dot{\theta} = \frac{p_\theta}{mR^2}$$

PROBLEM 7.10

A small bead of mass m can slide without friction on a parabolic wire of equation $z = ax^2$ rotating with angular velocity ω as in Figure 7.9.

  a. Write the Lagrangian (similarly with the problem in Chapter 6).
  b. Write the Hamiltonian.
  c. Find the highest point $z_{max} = Z$ at which the bead can position itself if the initial velocity at the origin of the coordinates is v_0.

Note: This is the same problem text as Problem 6.7 but now should be solved using Hamilton's equations.

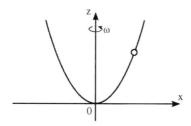

FIGURE 7.9 A bead is sliding without friction on a parabolic wire rotating about the vertical axis with angular velocity ω.

SOLUTION 7.10

a. Similarly with Chapter 6 (Problem 6.7), the Lagrangian is written by recalling that $z = ax^2$ and starting with the velocity as $v^2 = \dot{x}^2 + (x\omega)^2 + \dot{z}^2 = \dot{x}^2 + (x\omega)^2 + 4a^2x^2\dot{x}^2$.

The Lagrangian is $\mathcal{L} = T - U = \dfrac{m}{2}(\dot{x}^2 + x^2\omega^2 + 4a^2x^2\dot{x}^2) - mgax^2$.

b. The Hamiltonian is defined as

$$\mathcal{H} = \sum_{i=1}^{n} p_i \dot{q}_i - \mathcal{L}$$

Here, there is a one degree of freedom problem, so $\mathcal{H} = p\dot{q} - \mathcal{L}$, with $q = x$ and

$$p = \frac{\partial \mathcal{L}}{\partial \dot{x}} = m\dot{x} + 4ma^2x^2\dot{x}$$

Then the Hamiltonian is

$$\mathcal{H} = p\dot{q} - \mathcal{L} = m\dot{x}^2 + 4ma^2x^2\dot{x}^2 - \frac{m}{2}(\dot{x}^2 + x^2\omega^2 + 4a^2x^2\dot{x}^2) + mgax^2$$

$$= (\dot{x}^2 - x^2\omega^2 + 4a^2x^2\dot{x}^2) + mgax^2$$

Since the Lagrangian is time independent, the Hamiltonian is conserved and equal to the total energy, from the initial condition, at $x = 0, \dot{x} = v_0$, the Hamiltonian is written as

$$\frac{m}{2}(1 + 4a^2x^2)\dot{x}^2 - \frac{m}{2}(\omega^2 - 2ga)x^2 = \frac{mv_0^2}{2} \tag{7.4}$$

c. The bead moves to the highest point $z_{max} = Z$ given by the condition $\dot{x} = 0$. Equation (7.4) becomes

$$(2ga - \omega^2)x^2 = v_0^2$$

Multiplying on both sides by a and recalling that $z = ax^2$, it follows that the maximum height is

$$Z = \frac{av_0^2}{2ga - \omega^2}$$

It is easy to see that, in order to have a positive value for Z, the condition $2ga - \omega^2 > 0$ needs to be fulfilled, so $\omega^2 < 2ga$, or $\omega < \sqrt{2ga}$. In the case $\omega < \sqrt{2ga}$, the bead goes up to infinity on the parabola.

PROBLEM 7.11

Consider a bead of mass m on the spinning wire as in Figure 7.10. If the wire is in the shape of an upside-down Gaussian, specifically $f(\rho) = -\dfrac{1}{\sqrt{2\pi}} e^{-\frac{\rho^2}{2}}$, find any equilibrium positions and comment on their stability.

Note: This is the same problem text as Problem 6.10 but now should be solved using Hamilton's equations.

SOLUTION 7.11

In order to use Hamilton's equations, the Lagrangian is required. Recalling the position can be used to find velocity

$$\vec{r} = \rho\,\hat{\rho} + f(\rho)\hat{z} = \rho\,\hat{\rho} - \frac{1}{\sqrt{2\pi}} e^{-\frac{\rho^2}{2}}\,\hat{z}$$

with

$$\vec{v} = \dot{\rho}\,\hat{\rho} + \frac{1}{\sqrt{2\pi}}\rho\dot{\rho}e^{-\frac{\rho^2}{2}}\,\hat{z} + \rho\omega\hat{\phi}$$

the kinetic energy is given by

$$T = \frac{1}{2}mv^2 = \frac{1}{2}m\left(\dot{\rho}^2 + \frac{1}{2\pi}\rho^2\dot{\rho}^2 e^{-\rho^2} + \rho^2\omega^2\right)$$

and the potential energy is given by

$$U = mg\left(|f(0)| - |f(\rho)|\right) = mg\left(\frac{1}{\sqrt{2\pi}} - \frac{1}{\sqrt{2\pi}}e^{-\frac{\rho^2}{2}}\right) = \frac{mg}{\sqrt{2\pi}}\left(1 - e^{-\frac{\rho^2}{2}}\right)$$

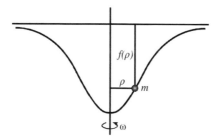

FIGURE 7.10   Bead on a spinning wire.

Therefore, the Lagrangian is

$$\mathcal{L} = T - U = \frac{1}{2} m \left(\dot{\rho}^2 + \frac{1}{2\pi} \rho^2 \dot{\rho}^2 e^{-\rho^2} + \rho^2 \omega^2 \right) - \frac{mg}{\sqrt{2\pi}} \left(1 - e^{-\frac{\rho^2}{2}} \right)$$

$$= \frac{1}{2} m \left(\dot{\rho}^2 \left(1 + \frac{1}{2\pi} \rho^2 e^{-\rho^2} \right) + \rho^2 \omega^2 \right) - \frac{mg}{\sqrt{2\pi}} \left(1 - e^{-\frac{\rho^2}{2}} \right)$$

To use this in Hamilton's equations, the velocity term must be substituted for the generalized momentum. Thus,

$$p = \frac{\partial \mathcal{L}}{\partial \dot{\rho}} = m \dot{\rho} \left(1 + \frac{1}{2\pi} \rho^2 e^{-\frac{\rho^2}{2}} \right)$$

$$\dot{\rho} = \frac{p}{m} \left(1 + \frac{1}{2\pi} \rho^2 e^{-\frac{\rho^2}{2}} \right)^{-1}$$

This can now be substituted into the Lagrangian

$$\mathcal{L} = \frac{1}{2} m \left(\frac{p^2}{m^2} \left(1 + \frac{1}{2\pi} \rho^2 e^{-\frac{\rho^2}{2}} \right)^{-2} \left(1 + \frac{1}{2\pi} \rho^2 e^{-\rho^2} \right) + \rho^2 \omega^2 \right) - \frac{mg}{\sqrt{2\pi}} \left(1 - e^{-\frac{\rho^2}{2}} \right)$$

$$= \frac{1}{2} m \left(\frac{p^2}{m^2} \left(1 + \frac{1}{2\pi} \rho^2 e^{-\frac{\rho^2}{2}} \right)^{-1} + \rho^2 \omega^2 \right) - \frac{mg}{\sqrt{2\pi}} \left(1 - e^{-\frac{\rho^2}{2}} \right)$$

$$= \frac{p^2}{2m} \left(1 + \frac{1}{2\pi} \rho^2 e^{-\frac{\rho^2}{2}} \right)^{-1} + \frac{\rho^2 \omega^2 m}{2} - \frac{mg}{\sqrt{2\pi}} \left(1 - e^{-\frac{\rho^2}{2}} \right)$$

The Hamiltonian can now be found via

$$\mathcal{H} = p \dot{\rho} - \mathcal{L}$$

$$\mathcal{H} = \frac{p^2}{m} \left(1 + \frac{1}{2\pi} \rho^2 e^{-\frac{\rho^2}{2}} \right)^{-1} - \frac{p^2}{2m} \left(1 + \frac{1}{2\pi} \rho^2 e^{-\frac{\rho^2}{2}} \right)^{-1} - \frac{\rho^2 \omega^2 m}{2} + \frac{mg}{\sqrt{2\pi}} \left(1 - e^{-\frac{\rho^2}{2}} \right)$$

$$= \frac{p^2}{2m}\left(1 + \frac{1}{2\pi}\rho^2 e^{-\frac{\rho^2}{2}}\right)^{-1} - \frac{\rho^2\omega^2 m}{2} + \frac{mg}{\sqrt{2\pi}}\left(1 - e^{-\frac{\rho^2}{2}}\right)$$

For stability analysis, consider Hamilton's equations

$$\dot{\rho} = \frac{\partial \mathcal{H}}{\partial p}$$

$$\dot{p} = -\frac{\partial \mathcal{H}}{\partial \rho}$$

Notice $\dot{\rho} = \dfrac{\partial \mathcal{H}}{\partial p}$ will not have ω in its expression. This means the p derivative is not useful for the analysis and will be ignored. Also, the ρ momentum must be zero so the first term in the Hamiltonian can be ignored as well. Therefore, consider

$$\dot{p} = -\frac{\partial \mathcal{H}}{\partial \rho}$$

$$0 = -\frac{\partial}{\partial \rho}\left(-\frac{\rho^2\omega^2 m}{2} + \frac{mg}{\sqrt{2\pi}}\left(1 - e^{-\frac{\rho^2}{2}}\right)\right)$$

$$0 = \rho\omega^2 m - \frac{mg}{\sqrt{2\pi}}\left(-e^{-\frac{\rho^2}{2}}(-\rho)\right)$$

$$\rho\omega^2 m = \frac{mg\rho}{\sqrt{2\pi}}e^{-\frac{\rho^2}{2}}$$

As was discussed previously, $\rho = 0$ is an uninteresting equilibrium point so it can be ignored. Therefore,

$$\omega = \sqrt{\frac{g}{\sqrt{2\pi}}}e^{-\frac{\rho^2}{4}}$$

which is exactly what was found in Chapter 6.

PROBLEM 7.12

Consider a mass m attached to a pendulum of length l and a spring of constant k as depicted in Figure 7.11. When the mass is at the lowest point of the pendulum, the spring is at its rest length. Considering a small displacement, find the angular velocity.

Note: This is the same problem text as Problem 6.11 but now should be solved using Hamilton's equations.

SOLUTION 7.12

In order to use Hamilton's equations, the Lagrangian is required. The kinetic energy of a pendulum should be familiar

$$T = \frac{1}{2}mv^2 = \frac{1}{2}ml^2\dot{\phi}^2$$

with the potential energy a combination of gravity and the spring

$$U = mgl(1-\cos\phi) + \frac{1}{2}k(l\sin\phi)^2$$

Considering small angles, the Lagrangian is given by

$$\mathcal{L} = T - U = \frac{1}{2}ml^2\dot{\phi}^2 - \frac{mgl}{2}\phi^2 - \frac{1}{2}kl^2\phi^2 = \frac{1}{2}ml^2\dot{\phi}^2 - \frac{l}{2}(mg - kl)\phi^2$$

To use this in Hamilton's equations, the velocity term must be substituted for the generalized momentum. Thus,

$$p = \frac{\partial \mathcal{L}}{\partial \dot{\phi}} = ml^2\dot{\phi}$$

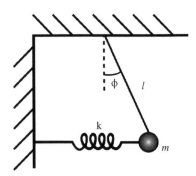

FIGURE 7.11 Mass connected to a pendulum and a spring.

$$\dot{\phi} = \frac{p}{ml^2}$$

Using this in the Lagrangian yields

$$\mathcal{L} = \frac{p^2}{2ml^2} - \frac{l}{2}(mg - kl)\phi^2$$

The Hamiltonian is then given by

$$\mathcal{H} = p\dot{\phi} - \mathcal{L} = \frac{p^2}{ml^2} - \frac{p^2}{2ml^2} + \frac{l}{2}(mg - kl)\phi^2$$
$$= \frac{p^2}{2ml^2} + \frac{l}{2}(mg - kl)\phi^2$$

Considering Hamilton's equations, the angular velocity can be found via

$$\dot{\phi} = \frac{\partial \mathcal{H}}{\partial p} = \frac{p}{ml^2}$$

$$\ddot{\phi} = \frac{\dot{p}}{ml^2}$$

where $\dot{p}$ comes from the other of Hamilton's equations

$$\dot{p} = -\frac{\partial \mathcal{H}}{\partial \phi} = -l(mg - kl)$$

Therefore,

$$\ddot{\phi} = \frac{\dot{p}}{ml^2} = -\frac{l(mg - kl)}{ml^2}\phi = -\frac{mg - kl}{ml}$$

and the angular velocity is given by

$$\omega = \sqrt{\frac{mg - kl}{ml}}$$

as was found in Chapter 6.

PROBLEM 7.13

A particle with initial mass m_0 and initial velocity v_0 begins losing mass according to the equation $m(t) = m_0 e^{-\alpha t}$ where α is constant. If there are no external forces, find an expression for the velocity.

Note: This is the same problem text as Problem 6.6 but now should be solved using Hamilton's equations.

SOLUTION 7.13

In order to use Hamilton's equations, the Lagrangian is required. The only energy in the problem is kinetic, so

$$\mathcal{L} = T = \frac{1}{2} m(t) \dot{x}^2 = \frac{1}{2} m_0 e^{-\alpha t} \dot{x}^2$$

To use this in Hamilton's equations, the velocity term must be substituted for the generalized momentum. So

$$p = \frac{\partial \mathcal{L}}{\partial \dot{x}} = m_0 e^{-\alpha t} \dot{x}$$

$$\dot{x} = \frac{p}{m_0} e^{\alpha t}$$

The Lagrangian can now be rewritten as

$$\mathcal{L} = \frac{1}{2} m_0 e^{-\alpha t} \left(\frac{p^2}{m_0^2} e^{2\alpha t} \right) = \frac{p^2}{2m_0} e^{\alpha t}$$

Therefore, the Hamiltonian is given by

$$\mathcal{H} = p\dot{x} - \mathcal{L} = \frac{p^2}{m_0} e^{\alpha t} - \frac{p^2}{2m_0} e^{\alpha t} = \frac{p^2}{2m_0} e^{\alpha t}$$

An expression for velocity can be found by taking Hamilton's equations

$$\dot{p} = -\frac{\partial \mathcal{H}}{\partial x} = 0$$

$$p = C$$

where C is a constant. Also,

$$\dot{x} = \frac{\partial \mathcal{H}}{\partial p} = \frac{p}{m_0} e^{\alpha t} = \frac{C}{m_0} e^{\alpha t}$$

Considering the initial velocity is v_0,

$$\dot{x}(0) = v_0 = \frac{C}{m_0}$$

$$C = m_0 v_0$$

and

$$\dot{x}(t) = v_0 e^{\alpha t}$$

exactly as was found before.

8 Coupled Oscillators and Normal Modes

8.1 THEORY

A variety of systems as atoms in molecules or molecules in certain types of polymers may oscillate as coupled oscillators, not independent oscillators. In this case, a different method shall be employed.

For a system of n degrees of freedom, the n generalized coordinates $q_1, q_2, \ldots, q_n$ can be written as a column matrix q. With $\mathcal{M}$ and $\mathcal{K}$ the mass and the spring constant matrices, respectively, the equation of motion for small oscillation about the equilibrium position (with $q = 0$) is

$$\mathcal{M}\ddot{q} = -\mathcal{K}q$$

The two matrices are obtained from the kinetic and potential energies of the system,

$$T = \frac{1}{2}\sum_{j,k} \mathcal{M}_{jk}\dot{q}_j\dot{q}_k$$

$$U = \frac{1}{2}\sum_{j,k} \mathcal{K}_{jk}q_jq_k$$

The motion in which all coordinates oscillate with the same frequency ω is called a normal mode and is written as

$$q(t) = \text{Re}(ae^{i\omega t})$$

The column matrix a satisfies the eigenvalue equation

$$(\mathcal{K} - \omega^2\mathcal{M})a = 0$$

If the matrix has a non-zero determinant, then the only solution is the trivial solution $a = 0$.

If the determinant is zero,

$$\det(\mathcal{K} - \omega^2\mathcal{M}) = 0$$

DOI: 10.1201/9781003365709-8

Then there is a non-trivial solution, and the frequencies ω at which the system is oscillating can be determined.

For a system with n degrees of freedom, there are n normal frequencies, each with a corresponding eigenvector.

8.2 PROBLEMS AND SOLUTIONS

PROBLEM 8.1

A system of oscillators with generalized coordinates q_1, q_2, q_3 has the following kinetic and potential energy (consider the masses and the spring constants equal to one):

$$T = \frac{1}{2}(\dot{q}_1^2 + \dot{q}_2^2 + \dot{q}_3^2)$$

$$U = \frac{1}{2}(4q_1^2 + 5q_2^2 + 4q_3^2 - 2q_1q_2 - 2q_2q_3)$$

a. Show that the position with $q_1 = q_2 = q_3 = 0$ is an equilibrium position.
b. Find the three normal frequencies.

SOLUTION 8.1

a. Condition for equilibrium

$$\frac{\partial U}{\partial q_i} = 0$$

For q_1,

$$\frac{\partial U}{\partial q_1} = 0$$

which yields to

$$4q_1 - q_2 = 0$$

For q_2,

$$\frac{\partial U}{\partial q_2} = 0$$

which yields to

$$5q_2 - q_1 - q_3 = 0$$

Lastly, for q_3,

$$\frac{\partial U}{\partial q_3} = 0$$

with the equation

$$4q_3 - q_2 = 0$$

The determinant of the system is

$$\det \begin{pmatrix} 4 & -1 & 0 \\ -1 & 5 & -1 \\ 0 & -1 & 4 \end{pmatrix} = 80 - 4 - 4 = 72 \neq 0$$

Since the determinant is not zero, the only solution is

$$q_1 = q_2 = q_3 = 0$$

Note that the potential energy can be rewritten as

$$U = \frac{1}{2}[3(q_1^2 + q_2^2 + q_3^2) + (q_1^2 - 2q_1q_2 + q_2^2) + (q_2^2 - 2q_2q_3 + q_3^2)]$$

$$= \frac{1}{2}[3(q_1^2 + q_2^2 + q_3^2) + (q_1 + q_2)^2 + (q_2 + q_3)^2]$$

It is easy to see that the total potential energy is positive, therefore the origin is a position of stable equilibrium.

b. The Lagrangian is

$$\mathcal{L} = T - U = \frac{1}{2}(\dot{q}_1^2 + \dot{q}_2^2 + \dot{q}_3^2) - \frac{1}{2}(4q_1^2 + 5q_2^2 + 4q_3^2 - 2q_1q_2 - 2q_2q_3)$$

Lagrange equations for q_1, q_2, q_3 are
   For q_1,

$$\frac{d}{dt}\frac{\partial \mathcal{L}}{\partial \dot{q}_1} = \frac{\partial \mathcal{L}}{\partial q_1}$$

$$\ddot{q}_1 = -4q_1 + q_2$$

For q_2,

$$\frac{d}{dt}\frac{\partial \mathcal{L}}{\partial \dot{q}_2} = \frac{\partial \mathcal{L}}{\partial q_2}$$

$$\ddot{q}_2 = -5q_2 + q_1 + q_3$$

For q_3,

$$\frac{d}{dt}\frac{\partial \mathcal{L}}{\partial \dot{q}_3} = \frac{\partial \mathcal{L}}{\partial q_3}$$

$$\ddot{q}_3 = -4q_3 + q_2$$

The three coupled equations can be written in the compact form as

$$\underbrace{\begin{bmatrix} 1 & 0 & 0 \\ 0 & 1 & 0 \\ 0 & 0 & 1 \end{bmatrix}}_{\mathcal{M}}\begin{bmatrix} \ddot{q}_1 \\ \ddot{q}_2 \\ \ddot{q}_3 \end{bmatrix} = -\underbrace{\begin{bmatrix} 4 & -1 & 0 \\ -1 & 5 & -1 \\ 0 & -1 & 4 \end{bmatrix}}_{\mathcal{K}}\begin{bmatrix} q_1 \\ q_2 \\ q_3 \end{bmatrix}$$

$$\det(\mathcal{K} - \omega^2\mathcal{M}) = 0$$

$$\det\left(\begin{bmatrix} 4 & -1 & 0 \\ -1 & 5 & -1 \\ 0 & -1 & 4 \end{bmatrix} - \omega^2\begin{bmatrix} 1 & 0 & 0 \\ 0 & 1 & 0 \\ 0 & 0 & 1 \end{bmatrix}\right) = 0$$

$$\det\left(\begin{bmatrix} 4-\omega^2 & -1 & 0 \\ -1 & 5-\omega^2 & -1 \\ 0 & -1 & 4-\omega^2 \end{bmatrix}\right) = 0$$

$$(4-\omega^2)(5-\omega^2)(4-\omega^2) - (4-\omega^2) - (4-\omega^2) = 0$$

which can be written as

$$(\omega^2-4)^2(\omega^2-5) - 2(\omega^2-4) = 0$$

or

$$(\omega^2-4)^2[(\omega^2-5)(\omega^2-4)-2] = 0$$

From here, $\omega_1 = 2$ (note that we choose the positive solutions only). The second equation becomes

$$\omega^4 - 9\omega^2 + 18 = 0$$

And the other two frequencies are calculated as $\omega_2 = \sqrt{6}$ and $\omega_3 = \sqrt{3}$.

PROBLEM 8.2

Consider a system of two masses and three springs, as pictured in Figure 8.1. For the following cases, find the two normal frequencies ω_1 and ω_2.

a. $k_1 = k_3 = k$; $k_2 = \dfrac{k}{3}$ and $m_1 = m_2 = m$;

b. $m_1 = 2m_2 = m$ and $k_1 = k_2 = k_3 = k$

SOLUTION 8.2

a. Using Newton's Second Law, $F = m\ddot{x}$, two equations $\mathcal{M}\ddot{x} = -\mathcal{K}x$ are obtained

$$m\ddot{x}_1 = -k_1 x_1 + k_2(x_2 - x_1) = -kx_1 + \frac{k}{3}(x_2 - x_1) = -\frac{4k}{3}x_1 + \frac{k}{3}x_2$$

$$m\ddot{x}_2 = -k_3 x_2 + k_2(x_1 - x_2) = -kx_2 + \frac{k}{3}(x_1 - x_2) = \frac{k}{3}x_1 - \left(\frac{k}{3} + k\right)x_2 = \frac{k}{3}x_1 - \frac{4}{3}x_2$$

The matrices $\mathcal{M}$ and $\mathcal{K}$ are found:

$$\mathcal{M} = \begin{bmatrix} m & 0 \\ 0 & m \end{bmatrix}$$

$$\mathcal{K} = \begin{bmatrix} \dfrac{4k}{3} & -\dfrac{k}{3} \\ -\dfrac{k}{3} & \dfrac{4k}{3} \end{bmatrix}$$

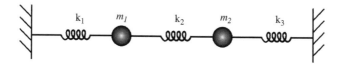

FIGURE 8.1 System of three springs and two masses.

From these, the matrix $\mathcal{K} - \omega^2 \mathcal{M}$ is written:

$$\mathcal{K} - \omega^2 \mathcal{M} = \begin{bmatrix} \dfrac{4k}{3} - m\omega^2 & -\dfrac{k}{3} \\[3mm] -\dfrac{k}{3} & \dfrac{4k}{3} - m\omega^2 \end{bmatrix}$$

Using $\det(\mathcal{K} - \omega^2 \mathcal{M}) = 0$, the two normal frequencies are determined

$$\left(\frac{4k}{3} - m\omega^2 \right) - \left(\frac{k}{3} \right)^2 = \left(\frac{5k}{3} - m\omega^2 \right)\left(\frac{2k}{3} - m\omega^2 \right)$$

$$\omega_1 = \sqrt{\frac{5k}{3m}}$$

$$\omega_2 = \sqrt{\frac{2k}{3m}}$$

b. The process is similar to part (a). Using Newton's Second Law, $F = m\ddot{x}$, two equations $\mathcal{M}\ddot{x} = -\mathcal{K}x$ are obtained

$$2m\ddot{x}_1 = -k_1 x_1 + k_2(x_2 - x_1) = -kx_1 + k(x_2 - x_1) = -2kx_1 + kx_2$$

$$m\ddot{x}_2 = -k_3 x_2 + k_2(x_1 - x_2) = -kx_2 + kx_1 - kx_2 = kx_1 - 2kx_2$$

The matrix $\mathcal{K} - \omega^2 \mathcal{M}$ is written:

$$\mathcal{K} - \omega^2 \mathcal{M} = \begin{bmatrix} 2k - 2m\omega^2 & -k \\ -k & 2k - m\omega^2 \end{bmatrix}$$

Using $\det(\mathcal{K} - \omega^2 \mathcal{M}) = 0$, the two normal frequencies are determined

$$(2k - 2m\omega^2)(2k - m\omega^2) - k^2 = 0$$

$$3k^2 - 64m\omega^2 + 2m^2\omega^4 = 0$$

Letting $W = \omega^2$, substitution is used to solve the equation

$$3k^2 - 64mW + 2m^2W^2 = 0$$

$$W = \frac{3+\sqrt{3}}{2m}k, -\frac{\sqrt{3}-3}{2m}k$$

So, the two normal frequencies are

$$\omega_1 = \sqrt{\frac{3+\sqrt{3}}{2m}k}$$

$$\omega_2 = \sqrt{-\frac{\sqrt{3}-3}{2m}k}$$

PROBLEM 8.3

Consider a system of two identical springs of spring constant k and two objects of mass m connected as in Figure 8.2. Ignoring the gravitational field, determine the frequencies ω at which the two objects oscillate.

SOLUTION 8.3

Considering y_1 and y_2 the positions of the two masses from the equilibrium positions, the kinetic energy is

$$T = \frac{m\dot{y}_1^2}{2} + \frac{m\dot{y}_2^2}{2}$$

and the potential energy is written as

FIGURE 8.2 A system of two objects of identical mass m connected by two identical springs of spring constant k.

$$U = \frac{ky_1^2}{2} + \frac{k(y_2 - y_1)^2}{2} = ky_1^2 + \frac{ky_2^2}{2} - ky_1y_2$$

The Lagrangian is

$$\mathcal{L} = T - U = \frac{m\dot{y}_1^2}{2} + \frac{m\dot{y}_2^2}{2} - ky_1^2 - \frac{ky_2^2}{2} + ky_1y_2$$

Lagrange equations for y_1 and y_2 become

$$\frac{d}{dt}\frac{\partial \mathcal{L}}{\partial \dot{y}_1} = \frac{\partial \mathcal{L}}{\partial y_1}$$

$$m\ddot{y}_1 = -2ky_1 + ky_2$$

$$\frac{d}{dt}\frac{\partial \mathcal{L}}{\partial \dot{y}_2} = \frac{\partial \mathcal{L}}{\partial y_2}$$

$$m\ddot{y}_2 = ky_1 - ky_2$$

Note that the two coupled equations can be written in the compact matrix form

$$\underbrace{\begin{bmatrix} m & 0 \\ 0 & m \end{bmatrix}}_{\mathcal{M}} \begin{bmatrix} \ddot{y}_1 \\ \ddot{y}_2 \end{bmatrix} = -\underbrace{\begin{bmatrix} 2k & -k \\ -k & k \end{bmatrix}}_{\mathcal{K}} \begin{bmatrix} y_1 \\ y_2 \end{bmatrix}$$

where the two square matrices are the mass matrix $\mathcal{M}$ and the spring constant matrix $\mathcal{K}$ with the column matrices $\ddot{y} = \begin{bmatrix} \ddot{y}_1 \\ \ddot{y}_2 \end{bmatrix}$ and $y = \begin{bmatrix} y_1 \\ y_2 \end{bmatrix}$

$$\mathcal{M}\ddot{y} = -\mathcal{K}y$$

To find the normal modes, consider the determinant

$$\det(\mathcal{K} - \omega^2\mathcal{M}) = 0$$

$$\det\left(\begin{bmatrix} 2k & -k \\ -k & k \end{bmatrix} - \omega^2 \begin{bmatrix} m & 0 \\ 0 & m \end{bmatrix}\right) = 0$$

$$\det\left(\begin{bmatrix} 2k - \omega^2m & -k \\ -k & k - \omega^2m \end{bmatrix}\right) = 0$$

$$2k^2 - 2km\omega^2 - km\omega^2 + m^2\omega^4 - k^2 = 0$$

$$k^2 - 3km\omega^2 + m^2\omega^4 = 0$$

After dividing by m^2 and by recalling that $\omega_e = \sqrt{\dfrac{k}{m}}$ the equation becomes

$$\omega^4 - 3\omega_e^2\omega^2 + \omega_e^4 = 0$$

$$\omega^2 = \frac{3\omega_e^2 \pm \sqrt{9\omega_e^4 - 4\omega_e^4}}{2} = \omega_e^2 \frac{3 \pm \sqrt{5}}{2}$$

$$\omega = \omega_e \sqrt{\frac{3 \pm \sqrt{5}}{2}}$$

These two normal frequencies are the frequencies at which the two spheres can oscillate in purely sinusoidal motion.

PROBLEM 8.4

Consider a system of three masses and three springs hanging vertically, as shown in Figure 8.3. All masses are equal, and all spring constants are equal. Find the matrix $K - \omega^2 M$.

SOLUTION 8.4

The kinetic energy of the system is

$$T = \frac{1}{2}m\dot{y}_1^2 + \frac{1}{2}m\dot{y}_2^2 + \frac{1}{2}m\dot{y}_3^2$$

The potential energy of the system is

$$U = \frac{1}{2}ky_1^2 + \frac{1}{2}k(y_2 - y_1)^2 + \frac{1}{2}k(y_3 - y_2)^2$$

$$= \frac{1}{2}ky_1^2 + \frac{1}{2}ky_2^2 + \frac{1}{2}ky_1^2 - ky_1y_2 + \frac{1}{2}ky_3^2 + \frac{1}{2}ky_2^2 - ky_2y_3$$

So the Lagrangian is

$$\mathcal{L} = T - U = \frac{1}{2}m\dot{y}_1^2 + \frac{1}{2}m\dot{y}_2^2 + \frac{1}{2}m\dot{y}_3^2 - \left(ky_1^2 + ky_2^2 - ky_1y_2 + \frac{1}{2}ky_3^2 - ky_2y_3\right)$$

FIGURE 8.3   System of three masses and three springs.

By derivation, a system of equations is obtained, $\mathcal{M}\ddot{y} = -\mathcal{K}y$.

$$\frac{d}{dt}\frac{\partial \mathcal{L}}{\partial \dot{y}_1} = \frac{\partial \mathcal{L}}{\partial y_1}$$

$$m\ddot{y}_1 = -2ky_1 + ky_2$$

$$\frac{d}{dt}\frac{\partial \mathcal{L}}{\partial \dot{y}_2} = \frac{\partial \mathcal{L}}{\partial y_2}$$

$$m\ddot{y}_2 = k(-2y_2 + y_1 + y_3)$$

$$\frac{d}{dt}\frac{\partial \mathcal{L}}{\partial \dot{y}_3} = \frac{\partial \mathcal{L}}{\partial y_3}$$

$$m\ddot{y}_3 = -ky_3 + ky_2$$

Thus, the matrix $\mathcal{M}$ is

$$\mathcal{M} = \begin{bmatrix} m & 0 & 0 \\ 0 & m & 0 \\ 0 & 0 & m \end{bmatrix}$$

And the matrix $\mathcal{K}$ is

$$\mathcal{K} = \begin{bmatrix} 2k & -k & 0 \\ -k & 2k & -k \\ 0 & -k & k \end{bmatrix}$$

So the matrix $\mathcal{K} - \omega^2 \mathcal{M}$ is the following:

$$\mathcal{K} - \omega^2 \mathcal{M} = \begin{bmatrix} -m\omega^2 + 2k & -k & 0 \\ -k & -m\omega^2 + 2k & -k \\ 0 & -k & -m\omega^2 + k \end{bmatrix}$$

PROBLEM 8.5

Consider two pendulums with strings of same length L, coupled by a spring of constant k as in Figure 8.4. Find the normal frequencies, if $m_1 = m_2 = m$, and in the small angle approximation.

SOLUTION 8.5

The kinetic energy of the system is

$$T = \frac{1}{2} mL^2 (\dot{\phi}_1^2 + \dot{\phi}_2^2)$$

The potential energy of the system is

$$U = mgL(1 - \cos\phi_1) + mgL(1 - \cos\phi_2) + \frac{1}{2}kL^2(\phi_2 - \phi_1)^2$$

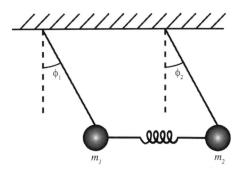

FIGURE 8.4 Coupled oscillators.

Using small angles, $\cos\phi_1$ is approximated:

$$\cos\phi_1 \approx 1 - \frac{1}{2}\phi_1^2$$

Rewriting the potential energy, the following expression is obtained:

$$U = \frac{mgL}{2}(\phi_1^2 + \phi_2^2) + \frac{1}{2}kL^2(\phi_2^2 + \phi_1^2 - 2\phi_1\phi_2)$$

The Lagrangian is

$$\mathcal{L} = T - U = \frac{1}{2}mL^2(\dot{\phi}_1^2 + \dot{\phi}_2^2) - \frac{mgL}{2}(\phi_1^2 + \phi_2^2) - \frac{1}{2}kL^2(\phi_2^2 + \phi_1^2 - 2\phi_1\phi_2)$$

By derivation, a system of equations is obtained, $\mathcal{M}\ddot{\phi} = \mathcal{K}\phi$.

$$\frac{d}{dt}\frac{\partial\mathcal{L}}{\partial\dot{\phi}_1} = \frac{\partial\mathcal{L}}{\partial\phi_1}$$

$$mL^2\ddot{\phi}_1 = -mgL\phi_1 - kL^2\phi_1 + kL^2\phi_2 = -(mgL + kL^2)\phi_1 + kL^2\phi_2$$

and

$$\frac{d}{dt}\frac{\partial\mathcal{L}}{\partial\dot{\phi}_2} = \frac{\partial\mathcal{L}}{\partial\phi_2}$$

$$mL^2\ddot{\phi}_2 = -mgL\phi_2 - kL^2\phi_2 + kL^2\phi_1 = kL^2\phi_1 - (mgL + kL^2)\phi_2$$

The matrices $\mathcal{M}$ and $\mathcal{K}$ are

$$\mathcal{M} = \begin{bmatrix} mL^2 & 0 \\ 0 & mL^2 \end{bmatrix}$$

$$\mathcal{K} = \begin{bmatrix} mgL + kL^2 & -kL^2 \\ -kL^2 & mgL + kL^2 \end{bmatrix}$$

The matrix $\mathcal{K} - \omega^2\mathcal{M}$ is written:

$$\mathcal{K} - \omega^2\mathcal{M} = \begin{bmatrix} mgL + kL^2 - \omega^2mL^2 & -kL^2 \\ -kL^2 & mgL + kL^2 - \omega^2mL^2 \end{bmatrix}$$

Using $\det(\mathcal{K} - \omega^2 \mathcal{M}) = 0$, the two normal frequencies are determined.

$$\det(\mathcal{K} - \omega^2 \mathcal{M}) = (mgL + kL^2 - \omega^2 mL^2)^2 - (kL^2)^2 = 0$$

$$(mgL + kL^2 - \omega^2 mL^2 - kL^2)(mgL + kL^2 - \omega^2 mL^2 + kL^2) = 0$$

The first normal frequency is

$$mgL - \omega^2 mL^2 = 0$$

$$\omega^2 = \frac{mgL}{mL^2} = \frac{g}{L}$$

$$\omega_1 = \sqrt{\frac{g}{L}}$$

The second normal frequency is

$$mgL + 2kL^2 - \omega^2 mL^2 = 0$$

$$\omega^2 = \frac{mgL + 2kL^2}{mL^2}$$

$$= \frac{mg + 2kL}{mL}$$

$$\omega_2 = \sqrt{\frac{g}{L} + \frac{2k}{m}}$$

PROBLEM 8.6

Consider the system of coupled pendulums as in Figure 8.5. If each pendulum has length L, spring constant k, and mass m, find the normal modes.

SOLUTION 8.6

Since the kinetic energy is the sum of energies due to the individual pendulums, it is given by

$$T = \frac{1}{2} mL^2 (\dot{\phi}_1^2 + \dot{\phi}_2^2)$$

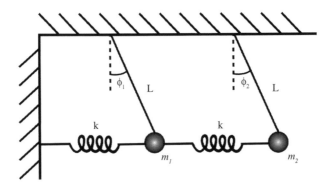

FIGURE 8.5 Two pendulums connected by springs.

The potential energy is due to the pendulums (gravitational) and the springs (elastic),

$$U_g = mgL(1-\cos\phi_1) + mgL(1-\cos\phi_2) \approx \frac{mgL}{2}(\phi_1^2 + \phi_2^2)$$

and

$$U_s = \frac{1}{2}k(L\sin\phi_1)^2 + \frac{1}{2}k(L\sin\phi_2 - L\sin\phi_1)^2$$

$$\approx \frac{1}{2}kL^2(\phi_1^2 + (\phi_2 - \phi_1)^2) = \frac{1}{2}kL^2(2\phi_1^2 - 2\phi_1\phi_2 + \phi_2^2)$$

where small angles were used to approximate the trigonometric functions. Therefore,

$$U = \frac{mgL}{2}(\phi_1^2 + \phi_2^2) + \frac{1}{2}kL^2(2\phi_1^2 - 2\phi_1\phi_2 + \phi_2^2)$$

The Lagrangian is now given by

$$\mathcal{L} = T - U = \frac{1}{2}mL^2(\dot\phi_1^2 + \dot\phi_2^2) - \frac{mgL}{2}(\phi_1^2 + \phi_2^2) - \frac{1}{2}kL^2(2\phi_1^2 - 2\phi_1\phi_2 + \phi_2^2)$$

and the normal frequencies can be found by considering Lagrange's equations

$$\frac{d}{dt}\frac{\partial\mathcal{L}}{\partial\dot\phi_1} = \frac{\partial\mathcal{L}}{\partial\phi_1}$$

and

$$\frac{d}{dt}\frac{\partial\mathcal{L}}{\partial\dot\phi_2} = \frac{\partial\mathcal{L}}{\partial\phi_2}$$

Thus, Lagrange equation for ϕ_1

$$\frac{d}{dt}\frac{\partial \mathcal{L}}{\partial \dot{\phi}_1} = \frac{\partial \mathcal{L}}{\partial \phi_1}$$

yields the equation

$$\frac{d}{dt}(mL^2\dot{\phi}_1) = -(mgL\phi_1 + kL^2(2\phi_1 - \phi_2))$$

$$mL^2\ddot{\phi}_1 = -((mgL + 2kL^2)\phi_1 - kL^2\phi_2)$$

and Lagrange equation for ϕ_2

$$\frac{d}{dt}\frac{\partial \mathcal{L}}{\partial \dot{\phi}_2} = \frac{\partial \mathcal{L}}{\partial \phi_2}$$

yields

$$\frac{d}{dt}(mL^2\dot{\phi}_2) = -(mgL\phi_2 + kL^2(\phi_2 - \phi_1))$$

$$mL^2\ddot{\phi}_2 = -(-kL^2\phi_1 + (mgL + kL^2)\phi_2)$$

Rewriting these equations as a matrix equation yields

$$\underbrace{\begin{bmatrix} mL^2 & 0 \\ 0 & mL^2 \end{bmatrix}}_{\mathcal{M}} \begin{bmatrix} \ddot{\phi}_1 \\ \ddot{\phi}_1 \end{bmatrix} = -\underbrace{\begin{bmatrix} mgL + 2kL^2 & -kL^2 \\ -kL^2 & mgL + kL^2 \end{bmatrix}}_{\mathcal{K}} \begin{bmatrix} \phi_1 \\ \phi_2 \end{bmatrix}$$

Therefore, the normal frequencies can be found by considering the determinant

$$\det(\mathcal{K} - \omega^2\mathcal{M}) = 0$$

$$\det\left(\begin{bmatrix} mgL + 2kL^2 & -kL^2 \\ -kL^2 & mgL + kL^2 \end{bmatrix} - \omega^2 \begin{bmatrix} mL^2 & 0 \\ 0 & mL^2 \end{bmatrix}\right) = 0$$

$$\det\left(\begin{bmatrix} mgL + 2kL^2 - \omega^2 mL^2 & -kL^2 \\ -kL^2 & mgL + kL^2 - \omega^2 mL^2 \end{bmatrix}\right) = 0$$

$$(mgL + 2kL^2 - \omega^2 mL^2)(mgL + kL^2 - \omega^2 mL^2) - k^2 L^4 = 0$$

$$(mgL + 2kL^2)(mgL + kL^2) - \omega^2 mL^2(mgL + kL^2 + mgL + 2kL^2) + \omega^4 m^2 L^4 - k^2 L^4 = 0$$

$$(mgL + 2kL^2)(mgL + kL^2) - \omega^2 mL^2(2mgL + 3kL^2) + \omega^4 m^2 L^4 - k^2 L^4 = 0$$

$$\omega^4 m^2 L^4 - \omega^2 mL^2(2mgL + 3kL^2) + m^2 g^2 L^2 + 3mgL^3 k + 2k^2 L^4 - k^2 L^4 = 0$$

$$\omega^4 m^2 L^2 - \omega^2 m(2mgL + 3kL^2) + m^2 g^2 + 3mgLk + k^2 L^2 = 0$$

$$\omega^2 = \frac{m(2mgL + 3kL^2) \pm \sqrt{m^2(2mgL + 3kL^2)^2 - 4m^2 L^2(m^2 g^2 + 3mgLk + k^2 L^2)}}{2m^2 L^2}$$

$$\omega^2 = \frac{2mg + 3kL \pm \sqrt{(2mg + 3kL)^2 - 4(m^2 g^2 + 3mgLk + k^2 L^2)}}{2mL}$$

$$\omega^2 = \frac{2mg + 3kL \pm \sqrt{4m^2 g^2 + 12mgkL + 9k^2 L^2 - 4m^2 g^2 - 12mgLk - 4k^2 L^2}}{2mL}$$

$$\omega^2 = \frac{2mg + 3kL \pm \sqrt{5k^2 L^2}}{2mL}$$

$$\omega^2 = \frac{g}{L} + \frac{k}{m}\left(3 \pm \sqrt{5}\right)$$

$$\omega = \sqrt{\frac{g}{L} + \frac{k}{m}\left(3 \pm \sqrt{5}\right)}$$

Notice, when $\omega = \sqrt{\frac{g}{L} + \frac{k}{m}\left(3 + \sqrt{5}\right)}$ the pendulums move toward/away from each other and when $\omega = \sqrt{\frac{g}{L} + \frac{k}{m}\left(3 - \sqrt{5}\right)}$ the pendulums move together to the left/right.

PROBLEM 8.7
Consider the cart from Figure 8.6, which is attached to the wall with a spring and has pendulum hanging below it. For a spring of constant k, a cart of mass m, and a pendulum of length L and mass M, find the normal frequencies. Assume small ϕ.

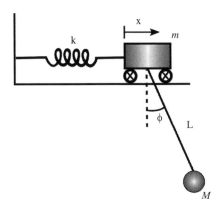

FIGURE 8.6   Cart connected to a spring with a pendulum connect beneath it.

SOLUTION 8.7

The normal frequencies can be found using the Lagrangian. In order to find expressions for energy the positions of the cart, $\vec{r}_m$, and pendulum, $\vec{r}_M$, must be found. These are given by

$$\vec{r}_m = x\,\hat{x}$$

$$\vec{r}_M = (x + L\sin\phi)\,\hat{x} - L\cos\phi\,\hat{y}$$

Considering small angles,

$$\vec{r}_M \approx (x + L\phi)\,\hat{x} - L\left(1 - \frac{\phi^2}{2}\right)\hat{y}$$

The velocities are

$$\vec{v}_m = \dot{x}\,\hat{x}$$

$$\vec{v}_M = (\dot{x} + L\dot{\phi})\,\hat{x} + L\phi\dot{\phi}\,\hat{y}$$

Therefore, the kinetic energy expressions are

$$T_m = \frac{1}{2}m\dot{x}^2$$

$$T_M = \frac{1}{2}M((\dot{x} + L\dot{\phi})^2 + (L\phi\dot{\phi})^2) = \frac{1}{2}M(\dot{x}^2 + 2L\dot{x}\dot{\phi} + L^2\dot{\phi}^2 + L^2\phi^2\dot{\phi}^2)$$

Since ϕ is small, $L^2\phi^2\dot\phi^2$ is very small and can be ignored. Thus,

$$T = T_m + T_M = \frac{1}{2}m\dot{x}^2 + \frac{1}{2}M(\dot{x}^2 + 2L\dot{x}\dot\phi + L^2\dot\phi^2) = \frac{1}{2}\dot{x}^2(m+M) + ML\left(\dot{x}\dot\phi + \frac{1}{2}L\dot\phi^2\right)$$

The potential energy has contributions from the spring and the pendulum. From the pendulum

$$U_M = MgL(1 - \cos\phi)$$

Considering small angles, this becomes

$$U_M \approx \frac{MgL}{2}\phi^2$$

The potential energy is then given by

$$U = \frac{1}{2}(kx^2 + MgL\phi^2)$$

The Lagrangian is then given by

$$\mathcal{L} = T - U = \frac{1}{2}\dot{x}^2(m+M) + ML\left(\dot{x}\dot\phi + \frac{1}{2}L\dot\phi^2\right) - \frac{1}{2}(kx^2 + MgL\phi^2)$$

In order to find the normal frequencies, Lagrange's equations must be considered:

$$\frac{d}{dt}\frac{\partial\mathcal{L}}{\partial\dot{x}} = \frac{\partial\mathcal{L}}{\partial x}$$

and

$$\frac{d}{dt}\frac{\partial\mathcal{L}}{\partial\dot\phi} = \frac{\partial\mathcal{L}}{\partial\phi}$$

Therefore, from the Lagrange equation in x,

$$\frac{d}{dt}((m+M)\dot{x} + ML\dot\phi) = -kx$$

$$(m+M)\ddot{x} + ML\ddot\phi = -kx$$

and from the Lagrange equation in ϕ:

$$\frac{d}{dt}(ML\dot{x} + ML^2\dot{\phi}) = -MgL\phi$$

$$ML\ddot{x} + ML^2\ddot{\phi} = -MgL\phi$$

$$\ddot{x} + L\ddot{\phi} = -g\phi$$

These can be re-expressed as the matrix equation

$$\underbrace{\begin{bmatrix} m+M & ML \\ 1 & L \end{bmatrix}}_{\mathcal{M}} \begin{bmatrix} \ddot{x} \\ \ddot{\phi} \end{bmatrix} = -\underbrace{\begin{bmatrix} k & 0 \\ 0 & g \end{bmatrix}}_{\mathcal{K}} \begin{bmatrix} x \\ \phi \end{bmatrix}$$

To find the normal modes, consider the determinant

$$\det(\mathcal{K} - \omega^2 \mathcal{M}) = 0$$

$$\det\left(\begin{bmatrix} k & 0 \\ 0 & g \end{bmatrix} - \omega^2 \begin{bmatrix} m+M & ML \\ 1 & L \end{bmatrix}\right) = 0$$

$$\det\left(\begin{bmatrix} k - \omega^2(m+M) & -\omega^2 ML \\ -\omega^2 & g - \omega^2 L \end{bmatrix}\right) = 0$$

By calculating the determinant, the following equation yields

$$(k - \omega^2(m+M))(g - \omega^2 L) - \omega^4 ML = 0$$

$$kg - (kL + g(m+M))\omega^2 + L(m+M)\omega^4 - ML\omega^4 = 0$$

$$kg - (kL + g(m+M))\omega^2 + mL\omega^4 = 0$$

From this equation, ω^2 is obtained as

$$\omega^2 = \frac{kL + g(m+M) \pm \sqrt{(kL + g(m+M))^2 - 4mLkg}}{2mL}$$

And by taking the radical, the frequencies of the normal modes are obtained

$$\omega = \sqrt{\frac{kL + g(m+M) \pm \sqrt{(kL + g(m+M))^2 - 4mLkg}}{2mL}}$$

Following a suggestion from Taylor, let $m = M = L = g = 1$ and $k = 2$ (in appropriate units). Therefore,

$$\omega = \sqrt{\frac{2 + 2 \pm \sqrt{(2+2)^2 - 4(2)}}{2}} = \sqrt{\frac{4 \pm \sqrt{8}}{2}} = \sqrt{2 \pm \sqrt{2}}$$

The two normal modes are then the cart and pendulum oscillating in the same direction, $\omega = \sqrt{2 - \sqrt{2}}$, and the cart and pendulum oscillating in opposite directions, $\omega = \sqrt{2 + \sqrt{2}}$.

PROBLEM 8.8

Consider the cart on a ramp (Figure 8.7), cart which is attached to the wall with a spring and has pendulum hanging below it. For a spring of constant k and an equilibrium position a distance l from the bottom of the ramp, a cart of mass m, a pendulum of length L and mass M, and a ramp at angle θ, find the normal frequencies. Assume small x and ϕ.

SOLUTION 8.8

Since the cart is on a ramp, the spring is already displaced from its equilibrium position, d, before the system oscillates. This distance is given by $\Sigma F_x = 0$

$$mg \sin \theta = kd$$

$$d = \frac{mg}{k} \sin \theta$$

Expressions for energy can now be found by considering the positions of the cart, $\vec{r}_m$, and the pendulum, $\vec{r}_M$

$$\vec{r}_m = x(\cos \theta \, \hat{x} - \sin \theta \, \hat{y})$$

with

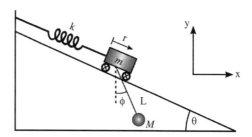

FIGURE 8.7   Cart on a ramp connected to a spring with a pendulum connect beneath it.

$$\vec{v}_m = \dot{\vec{r}} = \dot{x}(\cos\theta\,\hat{x} - \sin\theta\,\hat{y})$$

and

$$\vec{r}_M = (x\cos\theta + L\sin\phi)\hat{x} + (x\sin\theta - L\cos\phi)\hat{y} \approx (x\cos\theta + L\phi)\hat{x} + \left(x\sin\theta - \frac{L}{2}\phi^2\right)\hat{y}$$

where a small angle approximation was taken. Also

$$\vec{v}_M = (\dot{x}\cos\theta + L\dot{\phi})\hat{x} + (\dot{x}\sin\theta - L\phi\dot{\phi})\hat{y}$$

The energies can now be found. Starting with the cart, the kinetic energy is given by

$$T_m = \frac{1}{2}m\vec{v}_m \cdot \vec{v}_m = \frac{1}{2}m\dot{x}^2(\cos^2\theta + \sin^2\theta) = \frac{1}{2}m\dot{x}^2$$

and the potential energy is given by

$$U_m = mg(l - x\cos\theta) + \frac{1}{2}k(x+d)^2 = mg(l - x\cos\theta) + \frac{1}{2}k\left(x + \frac{mg}{k}\sin\theta\right)^2$$

Since the oscillations are small, the energy $U_m(x)$ can be expressed as

$$U_m = \frac{1}{2}\frac{\partial^2 U}{\partial x^2}x^2$$

Observe

$$\frac{\partial U}{\partial x} = -mg\cos\theta + k\left(x + \frac{mg}{k}\sin\theta\right)$$

and

$$\frac{\partial^2 U}{\partial x^2} = k$$

so

$$U_m = \frac{1}{2}kx^2$$

For the pendulum, the kinetic energy is

$$T_M = \frac{1}{2}M\vec{v}_M \cdot \vec{v}_M = \frac{1}{2}M((\dot{x}\cos\theta + L\dot{\phi})^2 + (\dot{x}\sin\theta - L\phi\dot{\phi})^2)$$

$$= \frac{1}{2}M(\dot{x}^2\cos^2\theta + 2\dot{x}\cos\theta L\dot{\phi} + L^2\dot{\phi}^2 + \dot{x}^2\sin^2\theta + 2\dot{x}\sin\theta L\phi\dot{\phi} + L^2\phi^2\dot{\phi}^2)$$

$$\approx \frac{1}{2}M(\dot{x}^2 + 2\dot{x}\cos\theta L\dot{\phi} + L^2\dot{\phi}^2)$$

where the last two terms in the parenthesis can be ignored; $\dot{x}\phi\dot{\phi}$ and $\phi^2\dot{\phi}^2$ are very small compared to the other terms. The potential energy is given by

$$U_M = Mg((l - x\cos\theta) + L(1 - \cos\theta)) \approx Mg\left(l - x\cos\theta + \frac{1}{2}L\phi^2\right)$$

Since the oscillations are small, $U_M(x,\phi)$ can be rewritten as

$$U_M = \frac{1}{2}\left(\frac{\partial^2 U}{\partial x^2}x^2 + 2\frac{\partial^2 U}{\partial x\partial\phi}x\phi + \frac{\partial^2 U}{\partial\phi^2}\phi^2\right)$$

with

$$\frac{\partial^2 U}{\partial x^2} = 0$$

$$\frac{\partial^2 U}{\partial x\partial\phi} = 0$$

$$\frac{\partial^2 U}{\partial\phi^2} = MgL$$

Therefore,

$$U_M = \frac{1}{2}MgL\phi^2$$

The total energies are then

$$T = T_m + T_M = \frac{1}{2}m\dot{x}^2 + \frac{1}{2}M(\dot{x}^2 + 2\dot{x}\cos\theta L\dot{\phi} + L^2\dot{\phi}^2)$$

$$= \frac{1}{2}(m + M)\dot{x}^2 + \dot{x}ML\dot{\phi}\cos\theta + \frac{1}{2}ML^2\dot{\phi}^2$$

and

$$U = U_m + U_M = \frac{1}{2}kx^2 + \frac{1}{2}MgL\phi^2$$

Putting everything together, the Lagrangian is given by

$$\mathcal{L} = T - U = \frac{1}{2}(m+M)\dot{x}^2 + \dot{x}ML\dot{\phi}\cos\theta + \frac{1}{2}ML^2\dot{\phi}^2 - \frac{1}{2}kx^2 - \frac{1}{2}MgL\phi^2$$

In order to find the normal frequencies, Lagrange's equations must be considered:

$$\frac{d}{dt}\frac{\partial\mathcal{L}}{\partial\dot{x}} = \frac{\partial\mathcal{L}}{\partial x}$$

and

$$\frac{d}{dt}\frac{\partial\mathcal{L}}{\partial\dot{\phi}} = \frac{\partial\mathcal{L}}{\partial\phi}$$

Therefore,

$$\frac{d}{dt}\frac{\partial\mathcal{L}}{\partial\dot{x}} = \frac{\partial\mathcal{L}}{\partial x}$$

$$\frac{d}{dt}((m+M)\dot{x} + ML\dot{\phi}\cos\theta) = -kx$$

$$(m+M)\ddot{x} + ML\ddot{\phi}\cos\theta = -kx$$

and

$$\frac{d}{dt}\frac{\partial\mathcal{L}}{\partial\dot{\phi}} = \frac{\partial\mathcal{L}}{\partial\phi}$$

$$\frac{d}{dt}(\dot{x}ML\cos\theta + ML^2\dot{\phi}) = -MgL\phi$$

$$\ddot{x}ML\cos\theta + ML^2\ddot{\phi} = -MgL\phi$$

Expressing this as a matrix equation yields

$$\underbrace{\begin{bmatrix} m+M & ML\cos\theta \\ ML\cos\theta & ML^2 \end{bmatrix}}_{\mathcal{M}}\begin{bmatrix} \ddot{x} \\ \ddot{\phi} \end{bmatrix} = -\underbrace{\begin{bmatrix} k & 0 \\ 0 & MgL \end{bmatrix}}_{\mathcal{K}}\begin{bmatrix} x \\ \phi \end{bmatrix}$$

and the normal modes can be found by considering the determinant.

$$\det(\mathcal{K} - \omega^2 \mathcal{M}) = 0$$

$$\det\left(\begin{bmatrix} k & 0 \\ 0 & MgL \end{bmatrix} - \omega^2 \begin{bmatrix} m+M & ML\cos\theta \\ ML\cos\theta & ML^2 \end{bmatrix}\right) = 0$$

$$\det\left(\begin{bmatrix} k - \omega^2(m+M) & -\omega^2 ML\cos\theta \\ -\omega^2 ML\cos\theta & MgL - \omega^2 ML^2 \end{bmatrix}\right) = 0$$

$$\omega^4 M^2 L^2 \cos^2\theta - (MgL - \omega^2 ML^2)(k - \omega^2(m+M)) = 0$$

$$\omega^4 M^2 L^2 \cos^2\theta - (kMgL - \omega^2((m+M)MgL + kML^2) + \omega^4(m+M)ML^2) = 0$$

$$\omega^4 ML \cos^2\theta - kg + \omega^2((m+M)g + kL) - \omega^4(m+M)L = 0$$

$$\omega^4(ML\cos^2\theta - mL - ML) + ((m+M)g + kL)\omega^2 - kg = 0$$

$$\omega^4(-mL - ML\sin^2\theta) + ((m+M)g + kL)\omega^2 - kg = 0$$

$$\omega^4(mL + ML\sin^2\theta) - ((m+M)g + kL)\omega^2 + kg = 0$$

$$\omega^2 = \frac{(m+M)g + kL \pm \sqrt{((m+M)g + kL)^2 - 4(mL + ML\sin^2\theta)(kg)}}{2(mL + ML\sin^2\theta)}$$

$$\omega = \sqrt{\frac{(m+M)g + kL \pm \sqrt{((m+M)g + kL)^2 - 4kg(mL + ML\sin^2\theta)}}{2(mL + ML\sin^2\theta)}}$$

Note: When $\theta = 0$, this reduces to what was found in the previous problem.

9 Nonlinear Dynamics and Chaos

9.1 THEORY

The systems encountered thus far have, for the most part, been linear in position, velocity, etc. While anything deviating from this may seem to a "special case", the reality is *most* systems are not so well behaved, that is, nonlinear. The simplest example of such a system is one which is governed by the logistics equation,

$$n_{t+1} = rn_t$$

where the "next value", n_{t+1}, is dependent on the "current value", n_t.

Nonlinear systems have interesting properties which require a full course in nonlinear dynamics to explore. This chapter serves as a brief sampling of this incredibly rich topic. In fact, the equations governing the system must be nonlinear to get *chaos* (however, not all nonlinear systems exhibit chaos). Trajectories of chaotic systems are not typically periodic, and these systems are very sensitive to initial conditions, that is, a slight difference in initial conditions can lead to drastically different trajectories. Numerical software is generally required to fully explore the chaos of a system and while numerically generated solutions are provided here, the authors recommend further education in numerical analysis to gain confidence in manipulating the equations describing such systems.

9.2 PROBLEMS AND SOLUTIONS

PROBLEM 9.1
Exponential population growth. By applying the simple example of a growth equation, calculate how many rabbits would populate Australia from 1850 to 1860, considering that two rabbits were brought to Australia and that the rabbits were reproducing at a rate of 20 rabbits per year (the rate could range between 18 and 30 rabbits per female rabbit).

SOLUTION 9.1
The number of rabbits in year $t+1$ is based on the number of rabbits in the previous year as

$$n_{t+1} = f(n_t) = rn_t$$

The function $f(n) = rn$, where r is the growth rate (here $r = 10$ per one female).

DOI: 10.1201/9781003365709-9

Starting with year 1850,

$$n_{1850+1} = f(n_{1850}) = rn_{1850}$$

$$n_{1850+2} = f(n_{1851}) = r^2 n_{1850}$$

And so on,

$$n_{1850+k} = f(n_{1850+k-1}) = r^k n_{1850}$$

And in ten years,

$$n_{1860} = n_{1850+10} = f(n_{1859}) = r^{10} n_{1850}$$

With our initial conditions $n_{1850} = 1$ and growth rate $r = 10$

$$n_{1860} = r^{10} n_{1850} = 10^{10} \cdot 1 = 10^{10}$$

After this model, in ten years the rabbit population increased to ten billion rabbits. The rabbits determined overgrazing, leading to decrease in plant diversity and land degradation and erosion. The migration of rabbits is thought to reach at times 80 miles in one year.

PROBLEM 9.2
The logistic growth model. Let $N(t)$ be the population dependent on time, based on maximum population growth rate r. The limiting factor is the carrying capacity K, which represents the total population that the environment could support based on the existent resources. Then the rate of change of population over time is

$$\frac{dN}{dt} = \frac{rN(K-N)}{K}$$

Calculate the rate of population $x = \dfrac{N}{K}$, the fraction of population alive divided by the total population that the environment could support.

SOLUTION 9.2
Starting with the formula

$$\frac{dN}{dt} = \frac{rN(K-N)}{K}$$

After the substitution of rate of alive population by the $x = \dfrac{N}{K}$, and by considering that $\dfrac{dN}{dt} = K\dfrac{dx}{dt}$, the equation becomes

$$\frac{dx}{dt} = rx(1-x)$$

Integrating

$$\int_{x_0}^{x} \frac{dx'}{x'(1-x')} = r\int_{0}^{t} dt'$$

$$(\ln x' - \ln(1-x'))\Big|_{x_0}^{x} = rt$$

$$\ln \frac{x}{1-x}\Big|_{x_0}^{x} = rt$$

$$\ln \frac{x}{1-x} - \ln \frac{x_0}{1-x_0} = rt$$

$$\ln \frac{x(1-x_0)}{x_0(1-x)} = rt$$

$$e^{rt} = \frac{x(1-x_0)}{x_0(1-x)}$$

$$x_0 e^{rt} - x_0 x e^{rt} = x - xx_0$$

$$x[1 + x_0(e^{rt}-1)] = x_0 e^{rt}$$

$$x(t) = \frac{x_0 e^{rt}}{1+x_0(e^{rt}-1)} = \frac{1}{\frac{1}{x_0 e^{rt}}+1-e^{-rt}} = \frac{1}{1+\left(\frac{1}{x_0}-1\right)e^{-rt}}$$

The fraction of the population x can be calculated, and this is called a sigmoid curve.

When solved numerically, the method is to take small intervals steps (logistic map, Markov chain), following:

$$x_{n+1} = rx_n(1-x_n)$$

where x_{n+1} is the population fraction for the next generation, x_n is the population fraction for present generation, and r is the growth rate.

PROBLEM 9.3

Considering that the Feigenbaum relation

$$(y_{n+1} - y_n) = \frac{1}{\delta}(y_n - y_{n-1})$$

is correct, confirm that the limit is $y_c = 1.0829$.

SOLUTION 9.3

The Feigenbaum relation is

$$(y_{n+1} - y_n) = \frac{1}{\delta}(y_n - y_{n-1})$$

with $y_1 = 1.0663$, $y_2 = 1.0793$, and $\delta = 4.6692016$.
Rewriting the relation, the following is obtained:

$$y_{n+1} - y_n\left(1 + \frac{1}{\delta}\right) + \frac{1}{\delta}y_{n-1} = 0$$

Using the characteristic root technique for sequences, the roots can be found:

$$x^2 - x\left(1 - \frac{1}{\delta}\right) + \frac{1}{\delta} = 0$$

$$(x-1)\left(x - \frac{1}{\delta}\right) = 0$$

So the roots are $r_1 = 1$ and $r_2 = \frac{1}{\delta}$. The recurrence is of the form

$$y_n = C_1\left(\frac{1}{\delta}\right)^n + C_2 1^n = C_1\left(\frac{1}{\delta}\right)^n + C_2$$

where C_1, C_2 are constants.
Using the given values of y_1, y_2, and δ solving for C_1 and C_2 is possible

$$C_1\left(\frac{1}{\delta}\right) + C_2 = 1.0663$$

$$C_1\left(\frac{1}{\delta^2}\right) + C_2 = 1.0793$$

Thus, $C_2 = 1.0663 - C_1\left(\dfrac{1}{\delta}\right)$. Plugging this value in the second equation, solving for C_1 is easy.

$$C_1\left(\frac{1}{\delta^2}\right) + 1.0663 - C_1\left(\frac{1}{\delta}\right) = 1.0793$$

$$C_1\left(\frac{1}{\delta^2} - \frac{1}{\delta}\right) = 0.013$$

$$C_1 = \frac{0.013}{\left(\dfrac{1}{\delta^2} - \dfrac{1}{\delta}\right)} = -0.07724$$

$$C_2 = 1.0663 - \frac{0.013}{\left(\dfrac{1}{\delta^2} - \dfrac{1}{\delta}\right)}\left(\frac{1}{\delta}\right) = 1.0829$$

So $y_c = \lim_{n \to \infty} y_n = \lim_{n \to \infty}\left(C_1\left(\dfrac{1}{\delta}\right)^n + C_2\right) = C_2$. Thus, $y_c = 1.0829$.

PROBLEM 9.4

Consider a driven damped pendulum with small drive strength $\gamma \ll 1$. Find a solution to equation of motion, using small angle approximation.

SOLUTION 9.4

A driven damped pendulum is governed by the equation

$$\ddot{\phi} + 2\beta\dot{\phi} + \omega_0^2 \sin\phi = \gamma\omega_0^2 \cos\omega t$$

Using small angle approximation, $\sin\phi = \phi$:

$$\ddot{\phi} + 2\beta\dot{\phi} + \omega_0^2\phi = \gamma\omega_0^2 \cos\omega t$$

To solve the differential equation, the general solution ϕ_n and a particular solution ϕ_t need to be found.

$$\phi = \phi_n + \phi_t$$

First, ϕ_n is determined:

$$\ddot{\phi} + 2\beta\dot{\phi} + \omega_0^2\phi = 0$$

To solve the equation, it is rewritten as $x^2 + 2\beta x + \omega^2 x = 0$. The roots of the equation are

$$x = \pm\sqrt{\beta^2 - \omega^2} - \beta$$

The solution for ϕ_n is

$$\phi_n = C_1 e^{t(\sqrt{\beta^2 - \omega^2} - \beta)} + C_2 e^{t(-\sqrt{\beta^2 - \omega^2} - \beta)}$$

Now, ϕ_t is found:

$$\ddot{\phi} + 2\beta\dot{\phi} + \omega_0^2 \phi = \gamma\omega_0^2 \cos\omega t$$

The following ϕ_t is a solution to the differential equation:

$$\phi_t = \frac{\gamma\omega \sin\omega t}{2\beta}$$

Using small angle approximation, the equation of motion of a damped pendulum is

$$\phi = C_1 e^{t(\sqrt{\beta^2 - \omega^2} - \beta)} + C_2 e^{t(-\sqrt{\beta^2 - \omega^2} - \beta)} + \frac{\gamma\omega \sin\omega t}{2\beta}$$

PROBLEM 9.5

Numerically solve and graph, for $t = 0$ to 100, the nonlinear differential equation $\ddot{\phi} + 2\beta\dot{\phi} + \omega_0^2 \sin\phi = \gamma\omega_0^2 \cos\omega t$ for $\omega = 1$, $\omega_0 = 2\omega$, $2\beta = \omega/3$, $\phi(0) = \dot{\phi}(0) = 0$ and

a. $\gamma = 0.3$
b. $\gamma = 1.06$
c. $\gamma = 2$

SOLUTION 9.5

a. The plot is represented in Figure 9.1.
b. The plot is represented in Figure 9.2.
c. The plot is represented in Figure 9.3.

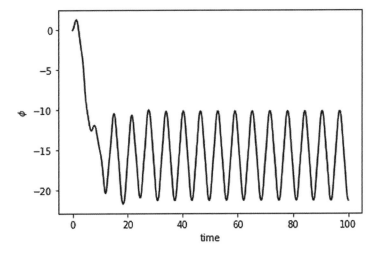

FIGURE 9.1   Driven damped oscillator with $\gamma = 0.3$.

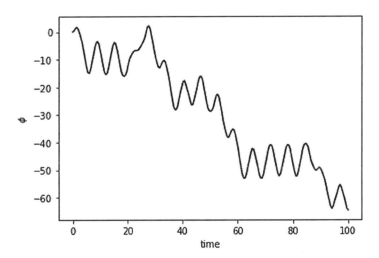

FIGURE 9.2   Driven damped oscillator with $\gamma = 1.06$.

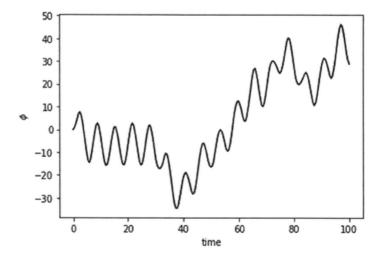

FIGURE 9.3   Driven damped oscillator with $\gamma = 2$.

PROBLEM 9.6

Consider the double pendulum in Figure 9.4 with masses m_1 and m_2, and lengths L_1 and L_2. Without assuming small angles, investigate the chaos of the system using numerical software.

SOLUTION 9.6

The equations of motion will be derived using Lagrange's equations. This requires the kinetic and potential energies. The kinetic energy can be found by considering the positions of the masses

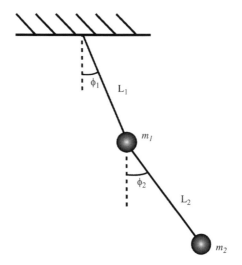

FIGURE 9.4   Double pendulum.

$$\vec{r}_1 = L_1(\sin\phi_1\,\hat{x} + \cos\phi_1\,\hat{y})$$

and

$$\vec{r}_2 = \vec{r}_1 + L_2(\sin\phi_2\,\hat{x} + \cos\phi_2\,\hat{y}) = (L_1\sin\phi_1 + L_2\sin\phi_2)\hat{x} + (L_1\cos\phi_1 + L_2\cos\phi_2)\hat{y}$$

The velocities are then given by

$$\vec{v}_1 = \dot{\vec{r}}_1 = L_1\dot{\phi}_1(\cos\phi_1\,\hat{x} - \sin\phi_1\,\hat{y})$$

and

$$\vec{v}_2 = \dot{\vec{r}}_2 = (L_1\dot{\phi}_1\cos\phi_1 + L_2\dot{\phi}_2\cos\phi_2)\hat{x} - (L_1\dot{\phi}_1\sin\phi_1 + L_2\dot{\phi}_2\sin\phi_2)\hat{y}$$

The kinetic energy of each mass is

$$T_1 = \frac{1}{2}m_1\vec{v}_1 \cdot \vec{v}_1 = \frac{1}{2}m_1 L_1^2\dot{\phi}_1^2$$

and

$$\begin{aligned}
T_2 &= \frac{1}{2}m_2\vec{v}_2 \cdot \vec{v}_2 = \frac{1}{2}m_2((L_1\dot{\phi}_1\cos\phi_1 + L_2\dot{\phi}_2\cos\phi_2)^2 + (L_1\dot{\phi}_1\sin\phi_1 + L_2\dot{\phi}_2\sin\phi_2)^2)\\
&= \frac{1}{2}m_2(L_1^2\dot{\phi}_1^2\cos^2\phi_1 + 2L_1L_2\dot{\phi}_1\dot{\phi}_2\cos\phi_1\cos\phi_2 + L_2^2\dot{\phi}_2^2\cos^2\phi_2 + L_1^2\dot{\phi}_1^2\sin^2\phi_1\\
&\quad + 2L_1L_2\dot{\phi}_1\dot{\phi}_2\sin\phi_1\sin\phi_2 + L_2^2\dot{\phi}_2^2\sin^2\phi_2)\\
&= \frac{1}{2}m_2(L_1^2\dot{\phi}_1^2 + 2L_1L_2\dot{\phi}_1\dot{\phi}_2(\sin\phi_1\sin\phi_2 + \cos\phi_1\cos\phi_2) + L_2^2\dot{\phi}_2^2)\\
&= \frac{1}{2}m_2(L_1^2\dot{\phi}_1^2 + 2L_1L_2\dot{\phi}_1\dot{\phi}_2\cos(\phi_1 - \phi_2) + L_2^2\dot{\phi}_2^2)
\end{aligned}$$

Therefore,

$$T = \frac{1}{2}(m_1 + m_2)L_1^2\dot{\phi}_1^2 + \frac{1}{2}m_2 L_2^2\dot{\phi}_2^2 + m_2 L_1 L_2\dot{\phi}_1\dot{\phi}_2\cos(\phi_1 - \phi_2)$$

The potential energies are given by

$$U_1 = m_1 g L_1(1 - \cos\phi_1)$$

and

$$U_2 = m_2 g(L_1(1 - \cos\phi_1) + L_2(1 - \cos\phi_2))$$

Therefore,

$$U = gL_1(m_1 + m_2)(1 - \cos\phi_1) + m_2 gL_2(1 - \cos\phi_2)$$

The Lagrangian is then given by

$$\mathcal{L} = T - U = \frac{1}{2}(m_1 + m_2)L_1^2\dot{\phi}_1^2 + \frac{1}{2}m_2 L_2^2\dot{\phi}_2^2 + m_2 L_1 L_2\dot{\phi}_1\dot{\phi}_2\cos(\phi_1 - \phi_2)$$
$$- gL_1(m_1 + m_2)(1 - \cos\phi_1) - m_2 gL_2(1 - \cos\phi_2)$$

The equations of motion are thus

$$\frac{d}{dt}\frac{\partial\mathcal{L}}{\partial\dot{\phi}_1} = \frac{\partial\mathcal{L}}{\partial\phi_1}$$

and

$$\frac{d}{dt}\frac{\partial\mathcal{L}}{\partial\dot{\phi}_2} = \frac{\partial\mathcal{L}}{\partial\phi_2}$$

Since the objective is to investigate chaos numerically, and for the sake of simplicity, all constants are set to 1. Therefore,

$$\mathcal{L} = \dot{\phi}_1^2 + \frac{1}{2}\dot{\phi}_2^2 + \dot{\phi}_1\dot{\phi}_2\cos(\phi_1 - \phi_2) - 2(1 - \cos\phi_1) - (1 - \cos\phi_2)$$
$$= \dot{\phi}_1^2 + \frac{1}{2}\dot{\phi}_2^2 + \dot{\phi}_1\dot{\phi}_2\cos(\phi_1 - \phi_2) - 3 + 2\cos\phi_1 + \cos\phi_2$$

with

$$\frac{d}{dt}\frac{\partial\mathcal{L}}{\partial\dot{\phi}_1} = \frac{\partial\mathcal{L}}{\partial\phi_1}$$

which yields

$$\sin(\phi_1 - \phi_2)\dot{\phi}_2^2 + 2(\sin\phi_1 + \ddot{\phi}_1) + \cos(\phi_1 - \phi_2)\ddot{\phi}_2 = 0$$

and

$$\frac{d}{dt}\frac{\partial\mathcal{L}}{\partial\dot{\phi}_2} = \frac{\partial\mathcal{L}}{\partial\phi_2}$$

which yields

$$\sin\phi_2 + \cos(\phi_1 - \phi_2)\ddot{\phi}_1^2 + \ddot{\phi}_2 - \sin(\phi_1 - \phi_2)\dot{\phi}_1^2 = 0$$

Since these differential equations involve trigonometric functions mixed with nonlinear second-order derivates, this cannot be solved analytically. Luckily, this is not required to investigate the chaos of the system; the differential equations can be solved numerically, given initial conditions. Consider a time range between 0 and 40, this system is solved for four different initial $(\phi_{1,i}, \phi_{2,i})$ pairs (in units of degrees): $(45, 45)$, $(46, 45)$, $(45, 46)$, and $(46, 46)$. Notice all these initial conditions are quite close. Plotting the results for ϕ_1 and ϕ_2 yields Figures 9.5 and 9.6.

Notice that despite the very close initial conditions, the trajectories of the pendulums are quite different; this is a hallmark of chaotic systems. Another way to look at this is the state-space orbit. Solving this system for $(\phi_{1,i}, \phi_{2,i}) = (45, 45)$ over the time range of 800 yields the following ϕ_1 versus $\dot{\phi}_1$ state-space orbit (Figure 9.7).

Observe the trajectory never repeats, another clear indication of chaos. Therefore, the double pendulum is clearly a chaotic system.

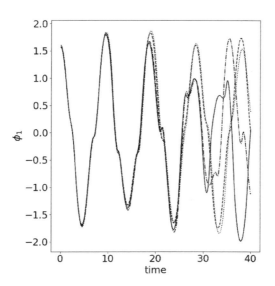

FIGURE 9.5 Plot of $\phi_1(t)$ for various initial positions.

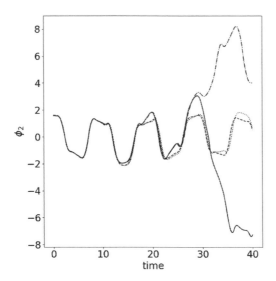

FIGURE 9.6   Plot of $\phi_2(t)$ for various initial positions.

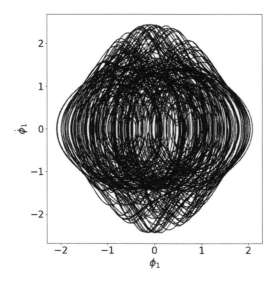

FIGURE 9.7   State-space plot of $\dot{\phi}_1(t)$ versus $\phi_1(t)$.

PROBLEM 9.7

Consider the two-pendulum system in Figure 9.8, each with mass m and length L, connected with a sprint of constant k. Without assuming small angles, investigate the chaos of the system using numerical software.

SOLUTION 9.7

The equations of motion will be obtained via Lagrange's equations. For this system, the kinetic energy is simply given by

$$T = \frac{1}{2}mL^2(\dot{\phi}_1^2 + \dot{\phi}_2^2)$$

and the potential energy is

$$U = mgL((1-\cos\phi_1) + (1-\cos\phi_2)) + \frac{1}{2}kL^2(\sin\phi_2 - \sin\phi_1)^2$$

$$= mgL(2 - \cos\phi_1 - \cos\phi_2) + \frac{1}{2}kL^2(\sin\phi_2 - \sin\phi_1)^2$$

where the last term in U is due to the spring. Therefore, the Lagrangian is given by

$$\mathcal{L} = T - U = \frac{1}{2}mL^2(\dot{\phi}_1^2 + \dot{\phi}_2^2) - mgL(2 - \cos\phi_1 - \cos\phi_2) - \frac{1}{2}kL^2(\sin\phi_2 - \sin\phi_1)^2$$

Since this will be analyzed numerically, the constants are set to 1. Thus,

$$\mathcal{L} = \frac{1}{2}(\dot{\phi}_1^2 + \dot{\phi}_2^2) - 2 + \cos\phi_1 + \cos\phi_2 - \frac{1}{2}(\sin\phi_2 - \sin\phi_1)^2$$

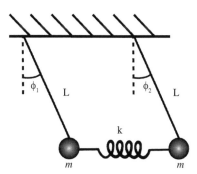

FIGURE 9.8 Two pendulums connected by a spring.

The equations of motion are thus

$$\frac{d}{dt}\frac{\partial \mathcal{L}}{\partial \dot{\phi}_1} = \frac{\partial \mathcal{L}}{\partial \phi_1}$$

and

$$\frac{d}{dt}\frac{\partial \mathcal{L}}{\partial \dot{\phi}_2} = \frac{\partial \mathcal{L}}{\partial \phi_2}$$

With the objective of investigating chaos numerically, numerical software can be used from this point out to calculate derivatives and plot results. Therefore, Lagrange equation for ϕ_1

$$\frac{d}{dt}\frac{\partial \mathcal{L}}{\partial \dot{\phi}_1} = \frac{\partial \mathcal{L}}{\partial \phi_1}$$

has the form

$$(1+\cos\phi_1)\sin\phi_1 + \ddot{\phi}_1 - \cos\phi_1\sin\phi_2 = 0$$

and Lagrange equation for ϕ_2

$$\frac{d}{dt}\frac{\partial \mathcal{L}}{\partial \dot{\phi}_2} = \frac{\partial \mathcal{L}}{\partial \phi_2}$$

has the form

$$\sin\phi_2 + \ddot{\phi}_2 - \cos\phi_2(\sin\phi_1 - \sin\phi_2) = 0$$

Given the mixture of derivatives and trigonometric functions, this cannot be solved analytically. Luckily, this is not required to investigate the chaos of the system; this can be solved numerical, given initial conditions. Consider a time range between 0 and 40, this system is solved for four different initial $(\phi_{1,i},\phi_{2,i})$ pairs (in units of degrees): $(25,45)$, $(26,45)$, $(25,46)$, and $(26,46)$. Notice all these initial conditions are quite close. Plotting the results for ϕ_1 and ϕ_2 yields Figures 9.9 and 9.10.

It appears the trajectories differ more and more each period, however this pattern does not clearly indicate chaos. To dig into this a bit more, the state-space orbit of ϕ_1 versus $\dot{\phi}_1$ can be calculated using the initial conditions $(\phi_{1,i},\phi_{2,i})=(25,45)$ over the time range of 800 (Figure 9.11).

While the individual trajectories did not show wildly diverging paths for very similar initial conditions, the state-space orbit clearly indicates this is a chaotic system. This suggests that the trajectories would have diverged more significantly given more time.

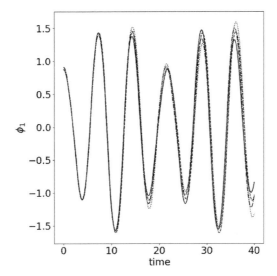

FIGURE 9.9 Plot of $\phi_1(t)$ for various initial positions.

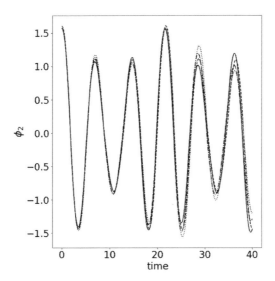

FIGURE 9.10 Plot of $\phi_2(t)$ for various initial positions.

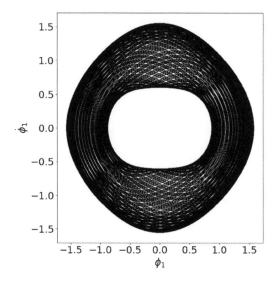

FIGURE 9.11   State-space plot of $\dot{\phi}_1(t)$ versus $\phi_1(t)$.

PROBLEM 9.8

Consider the cart below, which is attached to the wall with a spring and has pendulum hanging below it. For a spring of constant k, a cart of mass m, and a pendulum of length L and mass M, determine if this system is complicated enough to exhibit chaos. Do not assume small angles (Figure 9.12).

Note: This system was analyzed in Problem 8.7 using small angle approximation.

SOLUTION 9.8

The equations of motion will be obtained via Lagrange's equations. The kinetic energy of the cart is simply

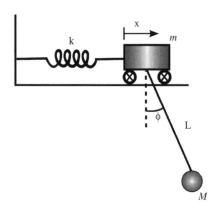

FIGURE 9.12   Cart connected to a spring with a pendulum oscillating beneath it.

$$T_m = \frac{1}{2}m\dot{x}^2$$

The position of the pendulum is

$$\vec{r}_M = (x + L\sin\phi)\hat{x} - L\cos\phi\,\hat{y}$$

so the velocity is

$$\vec{v}_M = (\dot{x} + L\dot{\phi}\cos\phi)\hat{x} + L\dot{\phi}\sin\phi\,\hat{y}$$

The kinetic energy of the pendulum is then given by

$$T_M = \frac{1}{2}M\vec{v}_M \cdot \vec{v}_M = \frac{1}{2}M(\dot{x}^2 + 2L\dot{x}\dot{\phi}\cos\phi + L^2\dot{\phi}^2)$$

and the total kinetic energy is

$$T = \frac{1}{2}(m+M)\dot{x}^2 + \frac{1}{2}ML^2\dot{\phi}^2 + ML\dot{x}\dot{\phi}\cos\phi$$

The potential is simply that of a spring and a pendulum.

$$U = \frac{1}{2}kx^2 + MgL(1-\cos\phi)$$

Therefore, the Lagrangian is

$$\mathcal{L} = T - U = \frac{1}{2}(m+M)\dot{x}^2 + \frac{1}{2}ML^2\dot{\phi}^2 + ML\dot{x}\dot{\phi}\cos\phi - \frac{1}{2}kx^2 - MgL(1-\cos\phi)$$

Since this will be analyzed numerically, the constants are set to 1, except M which is set to 2. Thus,

$$\mathcal{L} = \frac{3}{2}\dot{x}^2 + \dot{\phi}^2 + 2\dot{x}\dot{\phi}\cos\phi - \frac{1}{2}x^2 - 2 + 2\cos\phi$$

The equations of motion are thus

$$\frac{d}{dt}\frac{\partial\mathcal{L}}{\partial\dot{x}} = \frac{\partial\mathcal{L}}{\partial x}$$

and

$$\frac{d}{dt}\frac{\partial \mathcal{L}}{\partial \dot{\phi}} = \frac{\partial \mathcal{L}}{\partial \phi}$$

With the objective of investigating chaos numerically, numerical software can be used from this point out to calculate derivatives and plot results. Therefore, Lagrange equation for x

$$\frac{d}{dt}\frac{\partial \mathcal{L}}{\partial \dot{x}} = \frac{\partial \mathcal{L}}{\partial x}$$

yields to the equation

$$x + 3\ddot{x} + 2\ddot{\phi}\cos\phi - 2\dot{\phi}^2 \sin\phi = 0$$

and Lagrange equation for ϕ

$$\frac{d}{dt}\frac{\partial \mathcal{L}}{\partial \dot{\phi}} = \frac{\partial \mathcal{L}}{\partial \phi}$$

yields to the equation

$$\sin\phi + \ddot{x}\cos\phi + \ddot{\phi} = 0$$

Given the mixture of derivatives and trigonometric functions, this cannot be solved analytically. Luckily, this is not required to investigate the chaos of the system. Consider a time range between 0 and 40, this system is solved for three different initial (x_i, ϕ_i) pairs (angles are in degrees): $(2,45)$, $(2,46)$, and $(2,44)$. Notice all these initial conditions are quite close. Plotting the results for x and ϕ yields Figures 9.13 and 9.14.

There are clearly deviations of the trajectories which are larger than the initial conditions differences. This can be seen even more clearly by considering the state-space orbits x versus $\dot{x}$ and ϕ versus $\dot{\phi}$ over a time range of 800 and initial conditions $(x_i, \phi_i) = (2,45)$ (Figures 9.15 and 9.16).

It is clear these orbits do not repeat themselves indicating this system is in fact complex enough to exhibit chaos.

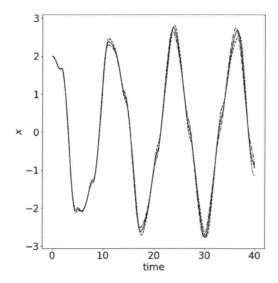

FIGURE 9.13   Plot of $x(t)$ for various initial positions.

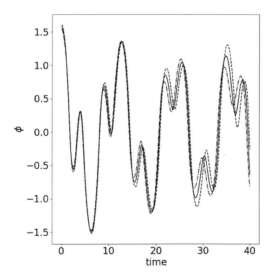

FIGURE 9.14   Plot of $\phi(t)$ for various initial positions.

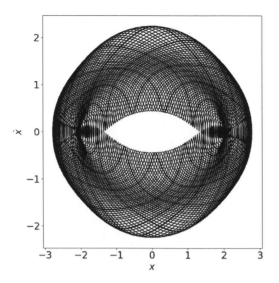

FIGURE 9.15   State-space plot of $\dot{x}(t)$ versus $x(t)$.

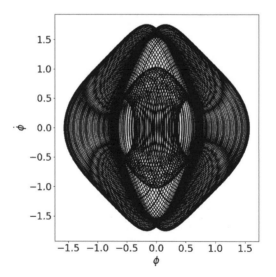

FIGURE 9.16   State-space plot of $\dot{\phi}(t)$ versus $\phi(t)$.

10 Special Relativity

10.1 THEORY

This chapter presents a brief discussion of special relativity and contains problems concerning length contraction, time dilation, as well as relativistic dynamics and redshift.

10.1.1 GALILEO'S TRANSFORMATIONS

Consider two reference frames S and S', with S' moving with a velocity V relative to S along x-axis. In Newtonian mechanics, it is an assumption that there is a universal time t. The origins of the two reference frames are chosen such that they coincide at time $t = 0$.

An event occurs in frame S at position $\vec{r} = (x, y, z)$ at time t, and it is measured in frame S' such that it occurs at $\vec{r}' = (x', y', z')$ at time t'. The following equations constitute Galilean transformations:

$$x' = x - Vt$$

$$y' = y$$

$$z' = z$$

$$t' = t$$

Furthermore, the vector position $\vec{r}'$ in frame S' can be written as $\vec{r}' = \vec{r} - \vec{V}t$ and the vector velocity $\vec{v}'$ in frame S' can be written as $\vec{v}' = \vec{v} - \vec{V}$, showing that in classical mechanics the velocity addition formula follows the vector addition rules (Figure 10.1).

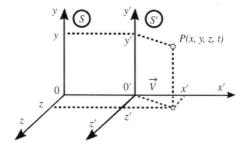

FIGURE 10.1 A system of two reference frames S and S', with frame S fixed and frame S' moving along the x direction with constant velocity.

DOI: 10.1201/9781003365709-10

Newton's Laws are invariant under the Galilean transformation.

Electromagnetism laws, however, are not invariant under the Galilean transformations. For example, Maxwell claims that electromagnetic wave propagates in vacuum with a speed $c = \dfrac{1}{\sqrt{\varepsilon_0 \mu_0}} = 3 \times 10^8$ m/s. If light is traveling with speed c with respect to reference frame S, and if the reference frame S' moves with a speed V along the x direction with respect to the reference frame S, then the speed in reference frame S' is given by $v' = c + V$, which would be greater than the speed of light. This obviously contradicts Maxwell's equations. This dilemma was solved by Einstein in his theory on special relativity.

An inertial reference frame is a reference frame (system of coordinates (x, y, z) and time t) in which all laws of physics hold.

10.1.2 POSTULATES OF THE THEORY OF RELATIVITY

First postulate of relativity: All laws of physics are the same in all inertial reference frames.

Second postulate of relativity: The speed of light in vacuum is the same in all inertial reference systems and in all directions.

In relativity, the time is not an invariant, as in Galilean transformations. Lorentz transformations are correct in relativistic mechanics and Maxwell's equations are invariant under Lorentz's transformations.

10.1.3 LORENTZ TRANSFORMATIONS

Consider two reference frames S and S', with S' moving with a velocity V relative to S along the x-axis (Figure 10.2). An event which occurs in frame S occurs at position $\vec{r} = (x, y, z)$ at time t, and measured in frame S' it occurs at $\vec{r}' = (x', y', z')$ at time t'. The following equations constitute Lorentz transformations:

$$x' = \frac{x - Vt}{\sqrt{1 - \dfrac{V^2}{c^2}}}$$

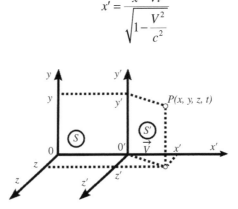

FIGURE 10.2 Two frames S and S', with frame S fixed and frame S' moving along the x direction with velocity $\vec{V}$.

$$y' = y$$

$$z' = z$$

$$t' = \frac{t - V\frac{x}{c^2}}{\sqrt{1 - \frac{V^2}{c^2}}}$$

Similarly, the equations for the coordinates in reference frame S as a function of coordinates in reference frame S' are

$$x = \frac{x' + Vt'}{\sqrt{1 - \frac{V^2}{c^2}}}$$

$$y = y'$$

$$z = z'$$

$$t = \frac{t' + V\frac{x'}{c^2}}{\sqrt{1 - \frac{V^2}{c^2}}}$$

Some usual notations are $\beta = \dfrac{V}{c}$ and Gamma factor is $\gamma = \dfrac{1}{\sqrt{1 - \beta^2}}$.

Note that Galileo's transformations are obtained for $c \to \infty$, or as approximation for $V \ll c$, $\beta = \dfrac{V}{c} \ll 1$.

All physics transformations are Lorentz transformation invariant, or L-invariant. Classical mechanics is invariant to Galileo's transformations, not to Lorentz transformations. For speeds close to speed of light, that is, in relativistic mechanics, Lorentz transformations need to be employed.

10.1.4 LENGTH CONTRACTION, TIME DILATION

The length l measured in the reference system in motion S' with velocity V in the x direction is smaller, compared to l_0, the length in the reference system at rest S (length contraction):

$$l = x_2' - x_1' = \left(x_2 \sqrt{1 - \frac{V^2}{c^2}} - Vt' \right) - \left(x_1 \sqrt{1 - \frac{V^2}{c^2}} - Vt' \right)$$

$$= (x_2 - x_1)\sqrt{1 - \frac{V^2}{c^2}} = l_0\sqrt{1 - \frac{V^2}{c^2}} = \frac{l_0}{\gamma} < l_0$$

The length of a ruler is maximum in the reference frame at rest.

Note: Only the longitudinal dimensions in the x direction (the direction of the velocity V) are changed, the dimensions in y and z directions are not changed. For example, a cube in reference frame S' becomes a parallelepiped with a smallest side in the x direction, and with sides in y and z directions equal to each other.

This is why the volume change is due to the change in dimension along the x direction, and has the formula indicated below. Please note that dV and dV_0 refer to the elements of volume in reference frames S' and S, while V refers to the velocity along the x direction.

$$dV = dV_0\sqrt{1 - \frac{V^2}{c^2}} = dV_0\sqrt{1 - \beta^2} = \frac{dV_0}{\gamma}$$

The time τ a process takes in a reference system in motion S' with velocity V in x direction is larger, compared to the time τ_0 it takes for the same process in the reference system at rest S (time dilation).

$$\tau = t_2' - t_1' = \frac{t_2 - \frac{Vx}{c^2}}{\sqrt{1 - \frac{V^2}{c^2}}} - \frac{t_1 - \frac{Vx}{c^2}}{\sqrt{1 - \frac{V^2}{c^2}}} = \frac{t_2 - t_1}{\sqrt{1 - \frac{V^2}{c^2}}} = \frac{\tau_0}{\sqrt{1 - \frac{V^2}{c^2}}} = \gamma \tau_0 > \tau_0$$

The time of a process observed from a reference frame in motion is larger compared to the time of a process observed from a reference frame at rest. The time of a process observed from the reference frame at rest is minimum.

On simultaneity: Simultaneity is relative, that is, two events simultaneous observed from a reference frame at rest S may not appear simultaneous while observed from a reference frame in motion S'.

The speed of light or the maximum speed of light in vacuum cannot be overcome. Speed of light in vacuum is the maximum attainable speed. For the limit speed V close to speed of light, the dimensions of the objects go to zero limit, and the periods of time move to an infinity limit.

10.1.5 COMPOSING VELOCITIES

By differentiating the Lorentz transformations, it yields that

$$dx = \frac{dx' + Vdt'}{\sqrt{1 - \frac{V^2}{c^2}}}$$

$$dy = dy' \qquad dz = dz'$$

$$dt = \frac{dt' + \dfrac{Vdx'}{c^2}}{\sqrt{1 - \dfrac{V^2}{c^2}}}$$

The velocities in reference frame S are obtained by dividing the first three equations by dt,

$$v_x = \frac{dx}{dt} = \frac{dx' + Vdt'}{dt' + \dfrac{Vdx'}{c^2}} = \frac{\dfrac{dx'}{dt'} + V}{1 + \dfrac{Vdx'}{c^2 dt'}} = \frac{v_x' + V}{1 + \dfrac{Vv_x'}{c^2}}$$

$$v_y = \frac{dy}{dt} = \frac{dy'\sqrt{1 - \dfrac{V^2}{c^2}}}{dt' + V\dfrac{dx'}{c^2}} = \frac{v_y'\sqrt{1 - \dfrac{V^2}{c^2}}}{1 + \dfrac{Vv_x'}{c^2}} = \frac{v_y'}{\gamma\left(1 + \dfrac{Vv_x'}{c^2}\right)}$$

$$v_z = \frac{dz}{dt} = \frac{dz'\sqrt{1 - \dfrac{V^2}{c^2}}}{dt' + V\dfrac{dx'}{c^2}} = \frac{v_z'\sqrt{1 - \dfrac{V^2}{c^2}}}{1 + \dfrac{Vv_x'}{c^2}} = \frac{v_z'}{\gamma\left(1 + \dfrac{Vv_x'}{c^2}\right)}$$

and the equivalent equations for the velocities in reference frame S'

$$v_x' = \frac{v_x - V}{1 - \dfrac{Vv_x}{c^2}}$$

$$v_y' = \frac{v_y\sqrt{1 - \dfrac{V^2}{c^2}}}{1 - \dfrac{Vv_x}{c^2}} = \frac{v_y}{\gamma\left(1 - \dfrac{Vv_x'}{c^2}\right)}$$

$$v_z' = \frac{v_z\sqrt{1 - \dfrac{V^2}{c^2}}}{1 - \dfrac{Vv_x'}{c^2}} = \frac{v_z}{\gamma\left(1 - \dfrac{Vv_x'}{c^2}\right)}$$

Minkowski space (spacetime) is a space with the three Euclidian special dimensions and one time dimension (x, y, z, ct). Note that the time t is multiplied by speed of light c for dimensional purposes.

10.1.6 RELATIVISTIC DYNAMICS

The relativistic (or variable) mass m_{rel} is dependent of velocity of the reference frame. If rest mass m is the mass measured in the reference frame at rest, the mass in a reference frame of the particle moving with velocity v in the x direction is

$$m_{rel} = m \frac{1}{\sqrt{1 - \dfrac{v^2}{c^2}}} = \frac{m}{\sqrt{1 - \beta^2}} = \gamma m$$

Linear momentum of a particle with rest mass m and velocity $\vec{v}$ is

$$\vec{p} = m_{rel}\vec{v} = m\vec{v} \frac{1}{\sqrt{1 - \dfrac{v^2}{c^2}}} = \frac{m\vec{v}}{\sqrt{1 - \beta^2}}$$

Note that when the speed v is close to speed of light, the momentum increases to infinity.

Relativistic kinetic energy K has the expression

$$K = E - E_0 = \frac{mc^2}{\sqrt{1 - \dfrac{v^2}{c^2}}} - mc^2 = (\gamma - 1)mc^2$$

The rest energy $E_0 = mc^2$.

The total energy of the particle is (Einstein equation)

$$E = m_{rel}c^2 = \frac{mc^2}{\sqrt{1 - \dfrac{v^2}{c^2}}} = \gamma mc^2$$

Relativistic linear momentum is related to the total energy of the particle.

$$E^2 = p^2c^2 + (mc^2)^2$$

Note: For photons, with $m = 0$, the equation becomes $E = pc$.

10.1.7 DOPPLER SHIFT

10.1.7.1 Redshift

When an object moves away from the Earth, the velocity of the object can be determined from the following equation, in which λ is the observed wavelength, λ_0 is the initial wavelength, and v is the speed of the galaxy, or object moving away (redshift) or closer (blueshift)

$$\lambda = \lambda_0 \frac{c+v}{c}$$

$$\frac{\Delta \lambda}{\lambda_0} = \frac{v}{c}$$

A redshift has been observed from different galaxies, which implied that they move away from each other, and therefore, the universe is expanding.

10.1.7.2 Blueshift

This has the same expression, but the velocity v has negative sign

$$\lambda = \lambda_0 \frac{c-v}{c}$$

$$\frac{\Delta \lambda}{\lambda_0} = -\frac{v}{c}$$

10.2 PROBLEMS AND SOLUTIONS

PROBLEM 10.1

An object is moving with velocity $\vec{v} = (v_x, 0, 0)$ with respect to the system S, and with velocity $\vec{v}' = (v_x', 0, 0)$ with respect to the system S', which is moving with a velocity $\vec{V} = (V, 0, 0)$ relative to S along the x-axis. Compose the speeds in both relativistic and classical way, and show that, in classical view, the absolute speed becomes greater than speed of light c. Consider $v_x' = V = \frac{3}{4}c$.

SOLUTION 10.1

Relativistic calculation

$$v_x = \frac{v_x' + V}{1 + \dfrac{V v_x'}{c^2}} = \frac{\dfrac{3c}{4} + \dfrac{3c}{4}}{1 + \dfrac{3c}{4}\dfrac{3c}{4}\dfrac{1}{c^2}} = \frac{\dfrac{6c}{4}}{1 + \dfrac{9}{16}} = \frac{6c}{4}\frac{16}{25} = \frac{24}{25}c < c$$

Classically,

$$v_x = v_x' + V = \frac{3}{4}c + \frac{3}{4}c = \frac{6}{4}c = \frac{3}{2}c > c$$

Note: This large speed requires, as expected, relativistic calculations, and this is why the classical method led to a speed greater than the speed of light, which is contradicting the relativistic expectation that the speed of light is the upper limit.

PROBLEM 10.2
A trip taking two years in intergalactic space takes one year observed from the Earth. Assuming the rocket moves in the x direction, find the speed V.

SOLUTION 10.2
The time is dilated in the reference frame in motion,

$$\tau = t_2' - t_1' = \frac{t_2 - t_1}{\sqrt{1 - \dfrac{V^2}{c^2}}} = \frac{\tau_0}{\sqrt{1 - \dfrac{V^2}{c^2}}}$$

so

$$\frac{\tau}{\tau_0} = \frac{1}{\sqrt{1 - \dfrac{V^2}{c^2}}}$$

By taking the square and rearranging,

$$\frac{\tau_0^2}{\tau^2} = 1 - \frac{V^2}{c^2}$$

$$\frac{V^2}{c^2} = 1 - \frac{\tau_0^2}{\tau^2}$$

$$V = c\sqrt{1 - \frac{\tau_0^2}{\tau^2}}$$

$$V = c\sqrt{1 - \left(\frac{1\ \text{year}}{2\ \text{years}}\right)^2} = c\sqrt{1 - \frac{1}{4}} = \frac{\sqrt{3}c}{2} = 0.866\,c$$

PROBLEM 10.3

Muons are unstable elementary particles with a charge equal to the charge of an electron, but the mass about 207 times larger than the mass of an electron. They are created by cosmic radiation colliding with atoms in the atmosphere. The lifetime of stationary muons is 30 times smaller than the lifetime of moving muons. Calculate the velocity of the muons. (The slow-moving muons have a lifetime of about 2.2 $\propto s$.)

SOLUTION 10.3

As before,

$$\tau = \frac{\tau_0}{\sqrt{1 - \dfrac{V^2}{c^2}}}$$

Here,

$$\tau_0 = \frac{1}{30}\tau$$

Following the calculations as before:

$$V = c\sqrt{1 - \frac{\tau_0^2}{\tau^2}} = c\sqrt{1 - \left(\frac{1}{30}\right)^2} = \frac{\sqrt{900 - 1}}{30}c = \frac{\sqrt{899}}{30}c = 0.99944\,c$$

PROBLEM 10.4

The life span of a giant Pacific octopus may be of about five years (the common octopus lives one to two years).

  a. Find the life span of the octopus when measured by an observer moving at a speed of $0.9c$ relative to the octopus.
  b. If the speed is increased by 10%, what happens to the time interval in the reference frame in motion compared to the one from point (a)?

SOLUTION 10.4

  a. The proper time interval measured in the rest reference frame of the octopus is five years. The time interval measured in the frame moving with speed V is

$$\tau = \gamma\tau_0 = \frac{\tau_0}{\sqrt{1 - \dfrac{V^2}{c^2}}} = \frac{5\ \text{years}}{\sqrt{1 - \dfrac{(0.9c)^2}{c^2}}} = \frac{5\ \text{years}}{\sqrt{1 - (0.9)^2}} = \frac{5\ \text{years}}{0.436} = 11.46\ \text{years}$$

b. Now, the velocity is increased by 10%, which means that

$$V_2 = 0.9\,c + 10\%(0.9\,c) = 1.1 \times 0.9\,c = 0.99\,c$$

$$\tau_2 = \gamma_2 \tau_0 = \frac{5\ \text{years}}{\sqrt{1 - \dfrac{(0.99\,c)^2}{c^2}}} = \frac{5\ \text{years}}{\sqrt{1 - (0.99)^2}} = \frac{5\ \text{years}}{0.141} = 35.44\ \text{years}$$

The time interval increases as

$$\frac{\tau_2 - \tau}{\tau} = \frac{35.44\ \text{years} - 11.46\ \text{years}}{11.46\ \text{years}} = 2.09 = 209\%$$

The 10% increase in velocity leads to an increase of 209% in the dilated time interval.

PROBLEM 10.5

Twins Alpha and Beta do an experiment when they are 20 years old. Alpha remains on the Earth (reference frame), and Beta goes on a trip with a speed of $V = 0.87c$. If the proper time interval τ_0 is 10 years, calculate the relativistic time, assuming that Beta arrives at the destination and goes promptly back with the same speed. What is the age of the twin brothers when Beta arrives back to Earth?

SOLUTION 10.5

Using the formula for the time dilation, it follows that

$$\tau = \gamma \tau_0 = \frac{\tau_0}{\sqrt{1 - \dfrac{V^2}{c^2}}} = \frac{10\ \text{years}}{\sqrt{1 - \dfrac{(0.87c)^2}{c^2}}} = \frac{10\ \text{years}}{\sqrt{1 - (0.87)^2}} = \frac{10\ \text{years}}{0.5} \cong 20\ \text{years}$$

Therefore, now, Alpha is 40 years old, and Beta is only 30 years old. Also, note that there is no paradox in fact – Beta is moving with respect to an observer at rest with respect to the Earth, while Alpha remains in the reference frame at rest with respect to the observer. Also, the positions of the twins cannot be interchanged, because Beta will feel that he is decelerating and stopping momentarily at the destination, and then promptly coming back with the same velocity V, while Alpha does not perceive any acceleration or deceleration in the reference frame at rest.

PROBLEM 10.6

A snake ("relativistic snake") is measured as having the length of one meter in a reference frame at rest with respect to the snake.

   a. Calculate the length of the snake in a reference frame traveling on x direction (along the snake) with velocity $V = 0.95c$?
   b. What is the length of the snake if the reference frame would move perpendicular to the snake?

SOLUTION 10.6

   a. The length of the snake measured in the reference frame moving with velocity V is

$$l = \frac{l_0}{\gamma} = l_0\sqrt{1-\frac{V^2}{c^2}} = 1\,\text{m}\sqrt{1-\frac{(0.95c)^2}{c^2}} = 1\,\text{m}\sqrt{1-(0.95)^2} = 0.31\,\text{m}$$

   b. If the reference frame moves perpendicular to the snake, the length of the snake is observed to be still one meter, it is not changed. However, the diameter of the snake will be measured in the moving reference system and become smaller following the same formula

$$d = \frac{d_0}{\gamma} = d_0\sqrt{1-\frac{V^2}{c^2}} = 0.31\,d_0$$

PROBLEM 10.7

The energy of an electron in fast motion is four times the rest energy.

   a. Calculate the rest energy.
   b. Calculate the speed V of the electron.
   c. Calculate the kinetic energy of the electron in eV.
   d. Calculate the linear momentum of the electron in eV/c.

SOLUTION 10.7

   a. The rest energy is

$$E_0 = mc^2 = 9.1\times10^{-31}\,\text{kg}\left(3\times10^8\,\frac{\text{m}}{\text{s}}\right)^2 = 8.19\times10^{-14}\,\text{J}$$

$$= \frac{8.19\times10^{-14}\,\text{J}}{1.6\times10^{-19}\,\dfrac{\text{J}}{\text{eV}}} = 5.12\times10^5\,\text{eV} = 0.512\,\text{MeV}$$

b. The relativistic energy is four times the rest energy

$$E = 4E_0 = 4\,mc^2$$

The relativistic mass is

$$E = m_{rel}c^2 = \frac{mc^2}{\sqrt{1 - \dfrac{v^2}{c^2}}} = \gamma mc^2$$

$$\frac{mc^2}{\sqrt{1 - \dfrac{v^2}{c^2}}} = 4\,mc^2$$

From here,

$$\sqrt{1 - \frac{v^2}{c^2}} = \frac{1}{4}$$

$$1 - \frac{v^2}{c^2} = \frac{1}{16}$$

$$\frac{v^2}{c^2} = 1 - \frac{1}{16}$$

which leads to

$$v = \sqrt{1 - \frac{1}{16}}\,c = \sqrt{\frac{15}{16}}\,c = \frac{\sqrt{15}}{4}\,c = 0.97\,c$$

c. Kinetic energy K is, using point (a),

$$K = E - E_0 = 4\,mc^2 - mc^2 = 3mc^2 = 3 \times 0.512 \text{ MeV} = 1.536 \text{ MeV}$$

d. Linear momentum is

$$E^2 = p^2c^2 + (mc^2)^2 = (4mc^2)^2$$

$$p^2c^2 = 16(mc^2)^2 - (mc^2)^2 = 15(mc^2)^2$$

Simplifying by c^2 and taking the square root,

$$p = \sqrt{15}\,\frac{mc^2}{c} = \sqrt{15}\,\frac{0.512 \text{ MeV}}{c} = 1.98 \text{ MeV}/c$$

PROBLEM 10.8

The spectroscopic measurement of light at wavelength $\lambda = 656\,\text{nm}$ coming from the (fictional) galaxy X has a redshift of $\Delta\lambda = 15\,\text{nm}$. Find the recessional speed of the galaxy.

SOLUTION 10.8

The fractional redshift is given by

$$\frac{\Delta\lambda}{\lambda_0} = \frac{v}{c}$$

The recessional speed is

$$v = \frac{\Delta\lambda}{\lambda_0}c = \frac{15\,\text{nm}}{656\,\text{nm}}c = 0.0228\,c = 2.28\times3\times10^6\,\frac{\text{m}}{\text{s}} = 6.86\times10^6\,\frac{\text{m}}{\text{s}}$$

The galaxy X is moving away from Earth with a speed of 2.28% of speed of light. Note that for objects in the solar system, the wavelength shift may be very small, less than 10^{-3} of an Angstrom, and there is a need for a very sensible spectrometer.

PROBLEM 10.9

Knowing the expressions for energy $E = \gamma mc^2$ and for linear momentum $p = \gamma mv$, find the relationship between the energy and linear momentum.

SOLUTION 10.9

Method I

From energy and momentum, by dividing the two relationships it yields that $\dfrac{E}{p} = \dfrac{c^2}{v}$ and the speed is $v = \dfrac{pc^2}{E}$

By substituting $\gamma = \dfrac{1}{\sqrt{1-\dfrac{v^2}{c^2}}}$ in the energy relationship for linear momentum $p = \gamma mv$,

$$p^2 = \frac{m^2v^2}{1-\dfrac{v^2}{c^2}} = \frac{m^2\dfrac{p^2c^4}{E^2}}{1-\dfrac{p^2c^4}{E^2}} = \frac{m^2p^2c^4}{E^2-p^2c^2}$$

By cross-multiplying,

$$p^2(E^2 - p^2c^2) = m^2p^2c^4$$

After dividing both sides by the squared momentum (which is not zero), the relationship energy momentum is found as

$$E^2 = p^2c^2 + m^2c^4$$

Method II

Note that the difference can be calculated in a more elegant way:

$$E^2 - p^2c^2 = (\gamma mc^2)^2 - (\gamma mv)^2 c^2 = \gamma^2m^2c^4 - \gamma^2m^2v^2c^2$$

$$= \gamma^2m^2c^4\left(1 - \frac{v^2}{c^2}\right) = \gamma^2m^2c^4\frac{1}{\gamma^2} = m^2c^4$$

From this it follows that

$$E^2 = p^2c^2 + m^2c^4$$

Note that, when considering the photon, with mass $m = 0$, the energy of the photon is simply

$$E = pc$$

Appendix
Differential Equations

In Classical Mechanics, many problems come down to solving an equation of motion. These equations can be comprised of positions, velocities, and accelerations; noting the last two are simply derivates of position. Therefore, the equations of motion are typically *differential equations.*

This appendix serves as a refresher on how to solve a few different types of differential equations which are encountered in this book. This is by no means a complete representation of the field of differential equations and is more of an appendix which can be referred to as problems in the various chapters are worked. For a complete treatment of differential equations, the authors recommend the books *Fundamentals of Differential Equations and Boundary Value Problems* by Nagle, Saff, and Snider, or *Elementary Differential Equations and Boundary Value Problems* by Boyce and DiPrima.

SEPARABLE EQUATIONS

Consider the first-order ordinary differential equation (ODE) of the form

$$\frac{dy}{dx} = f(x, y)$$

where $f(x, y)$ is some general function of x and y. For an equation to be separable, the function f can be rewritten as

$$f(x, y) = \frac{g(x)}{h(y)}$$

Therefore, the ODE becomes

$$\frac{dy}{dx} = \frac{g(x)}{h(y)}$$

This can be solved by rewriting the equation as

$$h(y)dy = g(x)dx$$

and integrating

$$\int h(y)\, dy = \int g(x)\, dx$$

Note: Do not forget the constant of integration!

FIRST-ORDER EQUATIONS WITH AN INTEGRATING FACTOR

Consider the first-order ODE of the form

$$\frac{dy}{dx} + p(x)y = q(x)$$

where $p(x)$ and $q(x)$ are general functions of x. Such an equation can be solved by considering an integrating factor $\mu(x)$ given by

$$\mu(x) = \exp\left(\int p(x)\,dx\right)$$

To understand how this transforms the original ODE, consider multiplying the original equation by $\mu(x)$

$$\mu(x)\frac{dy}{dx} + \mu(x)p(x)y = \mu(x)q(x)$$

Noting that

$$\frac{d}{dx}\mu(x) = p(x)\mu(x)$$

the original ODE can be rewritten as

$$\frac{d}{dx}(\mu(x)y) = \mu(x)q(x)$$

Therefore, the solution is given by

$$\mu(x)y = \int \mu(x)q(x)\,dx$$

$$y = \frac{1}{\mu(x)}\int \mu(x)q(x)\,dx$$

Note: Do not forget the constant of integration!

SECOND-ORDER HOMOGENEOUS EQUATIONS

Consider the second-order homogeneous differential equation of the form

$$a\frac{d^2y}{dx^2} + b\frac{dy}{dx} + cy = 0$$

where a, b, and c are constants. This type of equation can be solved by considering a solution of the form

$$y = e^{rx}$$

Note the following derivatives

$$\frac{dy}{dx} = re^{rx} = ry$$

$$\frac{d^2y}{dx^2} = r^2e^{rx} = r^2y$$

Substituting these into the original equation yields

$$ar^2y + bry + cy = 0$$

and since $y \neq 0$, this reduces to

$$ar^2 + br + c = 0$$

Therefore, finding the solution to the second-order homogeneous ODE amounts to finding the values of r that satisfy the above equation. Specially,

$$r_{\pm} = \frac{-b \pm \sqrt{b^2 - 4ac}}{2a}$$

Since there are two solutions for r, there are two solutions to the ODE. Therefore, the full solution is a linear combination of all solutions, specifically

$$y = c_1 e^{r_+ x} + c_2 e^{r_- x}$$

where c_1, c_2 are constants which can be found via the initial conditions of the problem.

Bibliography

Burlacu, Lucian, *Culegere de probleme de mecanica analitica*, David, Dorin Gheorghe, Universitatea din Bucuresti, Facultatea de Fizica, Bucuresti, Romania, 1988.

Byron, Frederick, *Mathematics of Classical and Quantum Physics*, Robert Fuller, Dover Publication, Inc., New York, 1992.

Goldstein, Herbert, *Classical Mechanics*, Charles Poole, John Safko, Addison Wesley, San Francisco, CA, 2002.

Hristev, Anatolie, *Mecanica si Acustica*, Editura Didactica si Pedagogica, Bucuresti, 1984.

Morin, David, *Introduction to Classical Mechanics: With Problems and Solutions*, 1st edition, Cambridge University Press, Cambridge, UK, 2008.

Nagle, Kent, Edward Saff, David Snider, *Fundamentals of Differential Equations and Boundary Value Problems*, Addison WesleyPearson, Boston, 2012.

Sauer, Timothy, *Numerical Analysis*, Pearson, Boston, 2012.

Taylor, John R., *Classical Mechanics*, University Science Books, Sausalito, CA, 2005.

Index

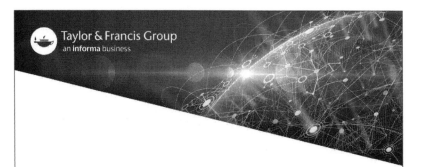

Printed and bound by CPI Group (UK) Ltd, Croydon, CR0 4YY

24/10/2024

01778280-0002